10th International Conference on Asian Energy Cooperation (AEC 2017)

E3S Web of Conferences Volume 27 (2018)

Irkutsk, Russia
29 - 31 August 2017

Editor:

Nikolai I. Voropai

ISBN: 978-1-5108-5569-4

Printed from e-media with permission by:

Curran Associates, Inc.
57 Morehouse Lane
Red Hook, NY 12571

Some format issues inherent in the e-media version may also appear in this print version.

Printed by Curran Associates, Inc. (2018)

For additional information, please contact EDP Sciences – Web of Conferences at the address below.

EDP Sciences – Web of Conferences
17, Avenue du Hoggar
Parc d'Activité de Courtabœuf
BP 112
F-91944 Les Ulis Cedex A
France

Phone: +33 (0) 1 69 18 75 75
Fax: +33 (0) 1 69 28 84 91

contact-edps@webofconferences.org

Additional copies of this publication are available from:

Curran Associates, Inc.
57 Morehouse Lane
Red Hook, NY 12571 USA
Phone: 845-758-0400
Fax: 845-758-2633
Email: curran@proceedings.com
Web: www.proceedings.com

TABLE OF CONTENTS

edp sciences

E3S Web of Conferences

Statement of Peer review

In submitting conference proceedings to E3S Web of Conferences, I certify to the Publisher that I adhere to the **Policy on Publishing Integrity** of the journal in order to safeguard good scientific practice in publishing.

1. All articles have been subjected to peer review administered by the proceedings editors.
2. Reviews have been conducted by expert referees, who have been requested to provide unbiased and constructive comments aimed, whenever possible, at improving the work.
3. Proceedings editors have taken all reasonable steps to ensure the quality of the materials they publish and their decision to accept or reject a paper for publication has been based only on the merits of the work and the relevance to the journal.

Title, date and place of the conference

10th International conference on Asian Energy Cooperation, August 29-31, 2017, Irkutsk, Russia

Proceedings editor(s):

Nikolai I. Voropai

Date and editor's signature December 05, 2017

E3S Web of Conferences **27**, 01001 (2018) https://doi.org/10.1051/e3sconf/20182701001
AEC 2017

Multilateral cooperation for power interconnection in Northeast Asia

Lyudmila Chudinova[1], *Sergei* Podkovalnikov[1,*], and *Ivan* Trofimov[1]

[1]Melentiev Energy Systems Institute of Siberian Branch of the Russian Academy of Sciences, Power systems Department, 130, Lermontov Str., Irkutsk, 664033, Russia

Abstract. Considered is experience of electric power integration in different parts of the world with emphasis on bilateral/multilateral interstate structures intended for promotion and development of interstate power grids. Analysed are Northeast Asian intergovernmental structures providing bilateral energy and power cooperation between Russia and other countries of the region. Multilateral cooperation in the field is just emerging, but should play a key role in promotion and development of Northeast Asian power grid.

Introduction

Electric power integration is a global process which has embraced many regions of the world. The interstate electric ties (ISETs) and power grids (ISPGs) have been developing in Europe, North and South America, some regions of Africa and Asia [1-10]. There are proposals on the creation of continental and intercontinental power grids: Eurasian, Trans-American, Pan-African, Asia-Pacific as well as Mediterranean, Black and Caspian seas power rings [1]. However, in some regions of the world power integration with the formation of the ISETs and ISPGs is still lagging behind for different reasons. These regions include in particular Northeast Asia (NEA).

A lot of studies of prospective ISETs and ISPG in NEA have been done [1,10-13, etc.]. Currently, transition from the stage of research and development to the stage of their implementation is nearing. It is required to form particular multilateral interstate/ intergovernmental structures in NEA to promote and develop megaproject of ISPG. International experience for South Asia and West Africa and existing bilateral and multilateral mechanisms for promoting ISETs and ISPG in NEA region are analyzed.

* Corresponding author: spodkovalnikov@isem.irk.ru

 https://doi.org/10.1051/e3sconf/20182701001

1 International experience

1.1 South Asian power cooperation

South Asia is a region characterized by high economic growth. Power demand growth remains robust due to high economic growth rates. The region has huge untapped hydropower potential. Most of the countries in South Asia are facing power shortages.

Power cooperation in the region has commenced in seventies when bilateral agreements between India and Nepal have been concluded to harness Nepal hydropower potential for mutual utilization. The first power cooperation bilateral agreements in South Asia were considered separately for different groups of countries and weren't coordinated within the entire region. South Asia Association of Regional Cooperation (SAARC) dedicated to improve the welfare of the people of South Asia through economic growth, social progress and cultural development was created in 1985. It comprises eight countries, including Afganistan, Pakistan, India, Bhutan, Nepal, Bangladesh, Sri Lanka, Maldives (Afganistan joined the organization in 2007). Fifteen years later, in 2000, Technical Committee on Energy was set up by SAARC. In 2006 SAARC Energy Center was created, and in 2009 SARC Energy Ministers approved the concept of a South Asia Energy Ring. Multilateral intergovernmental framework agreement on energy cooperation was signed among the member countries in 2014. These intergovernmental agreements realize conditions for meeting the ever-increasing demand of electricity for the socio-economic development of the region by means of cooperation in cross-border power exchange and trading through enhanced transmission interconnections, and grid connectivity would mutually benefit to all participant countries.

Current status of power system interconnection and proposed interstate links are presented in Fig.1 [14].

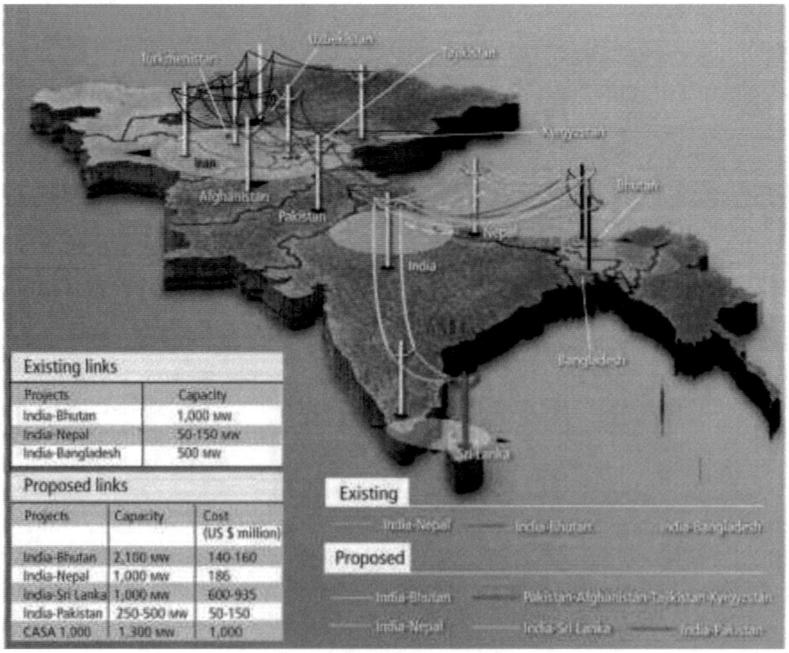

Fig. 1. South Asia power pool.

As can be seen from the above, power cooperation has been developing for decades and was moving upwards from low level of simple bilateral agreements to upper level of complicated multilateral arrangements. As result, SAARC institutional/organizational multilevel hierarchical system has been created to help promote and develop mega-project of regional-wide power interconnection. It is important, that power cooperation was developing under the auspices of regional-wide organization (SAARC), and this made easier overcoming barriers impeding power interconnection.

SAARC countries have been establishing transmission links among each other and with the countries outside the region, through which the trade of electricity realizes. The opportunities for Cross Border Electricity Trade (CBET) in the region has amplified energy safety and has become a political priority for all the state governments.

1.2 West African power cooperation

West African countries are among poorest countries in the world. Nonetheless they also pursue power cooperation policy which brings about system integration benefits including generating capacity and cost saving, reliability improvement, environmental impact reduction, etc.

In Africa, interconnections and cross-border electricity exchanges have been in existence for nearly half a century. Most of these interconnections originated from the development of some of the major hydropower projects. The West African Power Pool (WAPP) was formed in 2000 in taking due account of lessons learned from the establishment and development of the South African power pool.

Long before WAPP creation, Treaty on Economic Community of West Africa States (ECOWAS) was set up (in 1975), comprising fifteen countries (Benin, Burkina Faso, Cabo Verde , Côte D'ivoire, the Gambia, Ghana, Guinea, Guinea Bissau, Liberia, Mali, Niger, Nigeria, Senegal, Sierra Leone, Togo). Thus, WAPP was set up under the auspices of ECOWAS.

Current status of power system interconnection and proposed interstate links are given in Fig.2 [15].

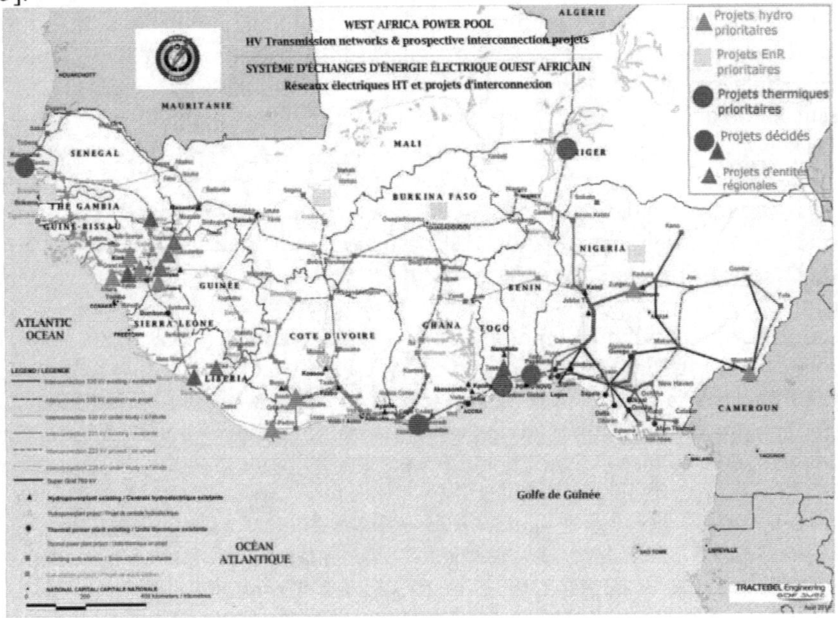

Fig. 2. West Africa power pool.

Soon after setting up WAPP, in 2001, ECOWAS Energy Ministers approved regional-wide master-plan which indicates that an estimated investment in the order of US$ 10 billion would be required for the construction of new power plants and upgrading and building new high voltage transmission lines over the next 15 years.

2 Northeast Asian power cooperation: Russia's case

Russia and NEA countries actively contact in power sector on the level of Governmental officials, bilateral intergovernmental commissions and working groups, industry and academic representatives. As a result of these activities multiple MOU and agreements signed. In the cases of Mongolia and China these activities resulted in construction of transboundary transmission lines and power trading.

Electric power trade between Russia and Mongolia has started in seventies years of twentieth century. Initially Russia-Mongolia's electric power cooperation was regulated within the framework of The Council for Mutual Economic Assistance. This Council comprised East European socialist countries, USSR, Mongolia and some others. Mongolia's power system has been operating in parallel with Unified Power System of USSR/Russia since those times. Mongolia was part of a large power system interconnection embracing countries mentioned above. This interconnection exceeded power system interconnection of West European countries in terms of installed generating capacity and power generation.

There exists Russia-Mongolia intergovernmental commission on trade, economic, scientific and technical cooperation, which also determines cooperation in power sector between countries. It prepares and issues documents which regulate activity of the countries in electric power.

Currently electric power cooperation and trading are based on Agreement between Governments of the countries on economic and near-border cooperation (1999). This Agreement is followed by Agreements between state owned power companies (between Russian and Mongolian System operators on parallel operation, dispatching and technical data exchange, 2008, and between Inter RAO EUS (Russia) and Power grid company of Mongolia on power supply, 2015). New Agreements concluded between both governmental and non-governmental companies (in particular, Eurosibenergo, Russia, and Just Group, Mongolia, 2010) consider development of Russia-Mongolia power trading and export to other countries, particularly to China.

As result of the bilateral activity, interstate transmission infrastructure with voltage up to 220 kV was created (Fig.3) [16,17]. There is power exchange between Mongolia and Russia (Fig.3). Russia mainly exports power to Mongolia, but in periods of minimum loads, Mongolia exports power back to Russia. This is due to Mongolia lacks peaking capacity which meets ever changing load, and existing Mongolian base-load thermal power plants have limited adjustment range for power output.

Russia and China have very intensive cooperative activity in power sector. There exists Russia-China intergovernmental commission on energy cooperation. Besides special Russia-China joint working group (WG) on cooperation in electric power is formed. Intergovernmental Agreement between these two countries on trade and economic cooperation, 1992, launched interstate power interconnection and trade. In this year 110 kV transmission line was put in operation and power export from Russia to China commenced, Fig.4 [16,17]. The Agreement was backed by the Intergovernmental Agreement on energy cooperation in 1996. Further, intergovernmental Agreements were followed by Agreements between power companies which now regulate technical details of power export from Russia to China (Agreement between RAO EES of Russia and China State grid corporation, 1999, 2005-2006; Agreement between Federal State company of Russia and China State grid corporation, 2010, 2014; Contract between East power company, Russia, and China State grid corporation, 2012; Agreement between Federal grid company and System operator, Russia, and Northeast China grid company, 2012).

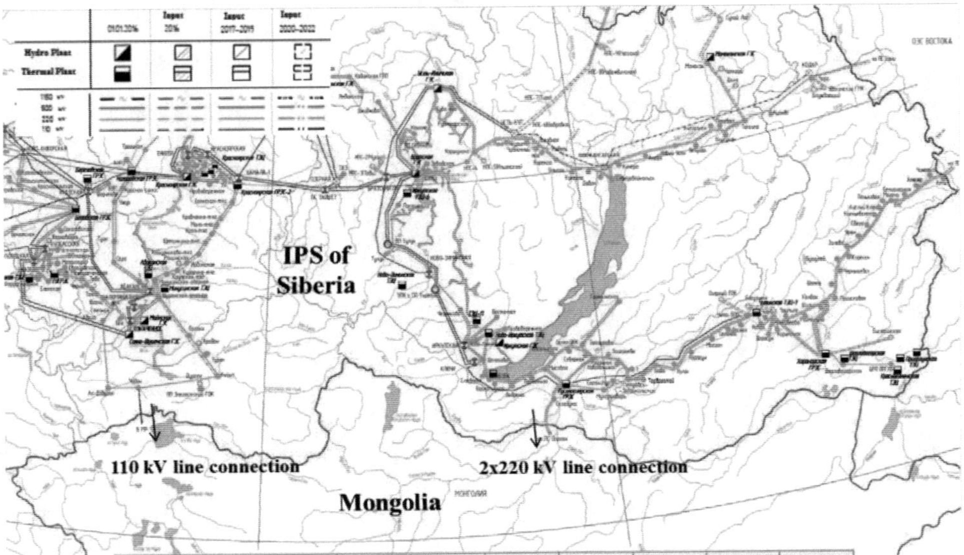

Years	2010	2011	2012	2013	2014	2015	2016
Export to Mongolia, GWh	214	263	393	414	390	284	300
Import from Mongolia, GWh	21	21	21	23	30	54	34

Fig. 3. Russia-Mongolia power interconnection (IPS – Interconnected Power System)

According to the above Agreements Russia-China interstate transmission infrastructure was expanded at the cost of 220 kV transmissions, and at last 500 kV transmission line was put in place in 2011 with back-to-back facility constructed on Chinese territory (Fig. 4). Sharp increase of power export from 2011, when line 500 kV was put in place, can be seen from Fig.4. As a result, power export to China reached about 3,5 TWh/year in 2013, and than was slightly reducing.

Wide scale export of power from Russia to China also has been considered starting since 1999. It was supposed to be realized in three stages. Upon completion of the third stage the export will reach 60 TWh/year. On the first stage export should be about 3,5-4 TWh/year. Thus, in terms of quantity the first stage has already been virtually realized.

Russian companies like Eurosibenergo, Synthesis, RusHydro, RusTech appear in the playground recent years. They conclude Agreements with Chinese counterparts on electric power cooperation, especially on export of power to China, during 2010-2014.

There exists Russia-Democratic People's Republic of Korea (DPRK) intergovernmental commission on trade, economic, scientific and technical cooperation, which determines also cooperation in power sector between countries. Additionally joint WG on cooperation in electric power is formed. Concerted activity in power sector between Russia and DPRK is held within the framework of the Commission and the WG.

Agreement between RAO UPS of Russia and Ministry of electric power and coal industry of DPRK has been concluded in 2001. It determined bilateral inter-country activity in electric power, particularly conducting study of Russia-DPRK interconnection with extension to Republic of Korea (RoK) and power trading among these countries. Later on (in 2015), Intergovernmental Agreement on cooperation in electric power was put in force. It provided contemporary basis for Russia-DPRK cooperation in this field.

https://doi.org/10.1051/e3sconf/20182701001

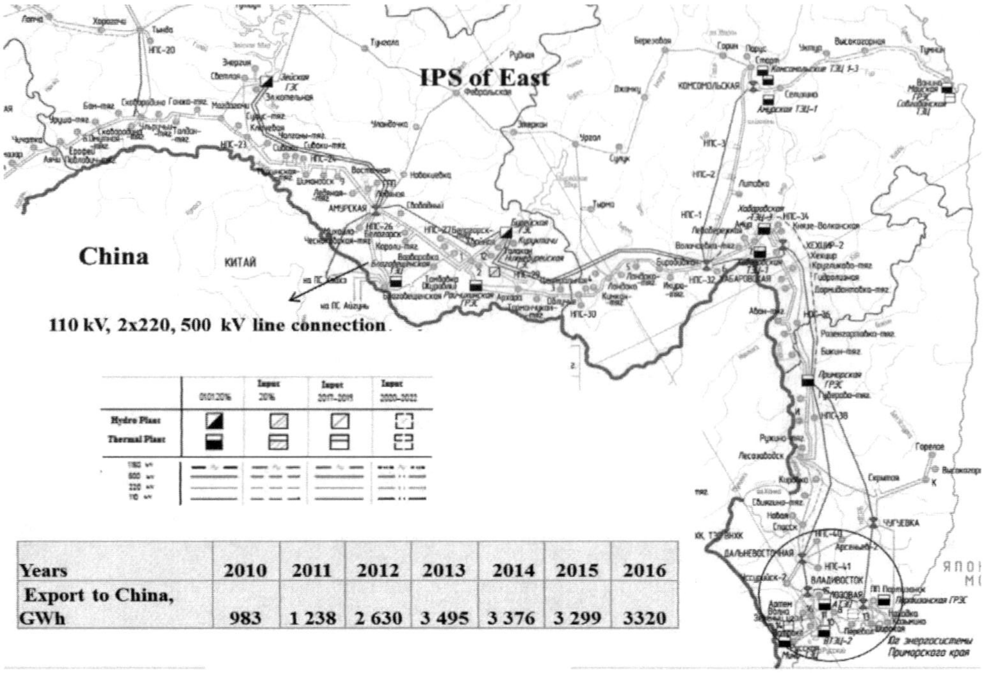

Years	2010	2011	2012	2013	2014	2015	2016
Export to China, GWh	983	1 238	2 630	3 495	3 376	3 299	3320

Fig. 4. Russia-China power interconnection.

Russia-DPRK interconnection proposal has long history since eighties. It was proposed to connect South of Russian Far East (RFE) and Northeast (NE) part of DPRK by alternating current (AC) transmission line. As far as DPRK power system operates at frequency of 60 Hz, NE part of DPRK power system should be disconnected from the rest DPRK power system and synchronized with Russian power system at frequency 50 Hz. Russia was supposed to export power to DPRK load island through this AC line. But after USSR dissolution the project was abolished, and it was renewed in post-soviet period, however its future is still uncertain.

Thus, DPRK demonstrated its intention to interconnection with USSR/Russia and even ROK. And this intention is present in the mentioned Agreements.

There also exists Russia-RoK intergovernmental commission on economic, scientific and technical cooperation. Legal basis for electric power cooperation provides Intergovernmental Agreement on cooperation in energy (2000). Memorandums of Understanding (MOUs) signed (2005, 2015) between Russian and ROK power companies on cooperation, in particular on potential power interconnections among countries.

Although there exists Russia-Japan intergovernmental commission on trade and economic issues. Russian and Japanese companies were actively discussing the potential projects of power interconnection between countries. At the turn of 2000 large study of Russia-Japan power bridge project was conducted. Only recently (2016) company's activity was backed by intergovernmental decision taken by Russia-Japan intergovernmental commission to study Russia-Japan power bridge project.

Just recently bilateral cooperation in NEA region advanced at the level of multilateral cooperation among industrial, governmental and international partners in NEA-wide scale. Particularly, signed in 2013 was MOU on Study of Gobitec and Asian Supergrid. It united Energy Economics Institute (RoK), Ministry of Energy (Mongolia), Renewable Energy Foundation (Japan), JSC "Institute of Energy Systems" (Russia) and International body, Energy Charter. One more MOU on Study of NEA-wide grid prospects was signed by

Russian grids company (Rosseti), Korea Electric Power Corporation (KEPCO), China State Grid Corporation, Soft Bank (Japan) in 2016.

Intergovernmental multilateral cooperation in NEA region falls behind so far.

3 Organizational and institutional system for development of NEA power interconnection

What was said above can be generalized in the following multilevel system (Fig.5). Now the system is developing from bottom to up - bottom-up approach is dominating. There exist two lower levels (governmental and non-governmental bilateral activities). The third level is just appearing (non-governmental multilateral activities), and the forth level (NEA-wide multilateral intergovernmental arrangements) is absent. It is needed to strengthening the third level by multilateral Agreements and creating the forth level – multilateral intergovernmental NEA-wide body.

As a result, the bottom – up approach will be complemented by top – down approach, because top-level NEA-wide intergovernmental body will generate signals which will stream downwards to country's level to shape concertedly all-region power interconnection.

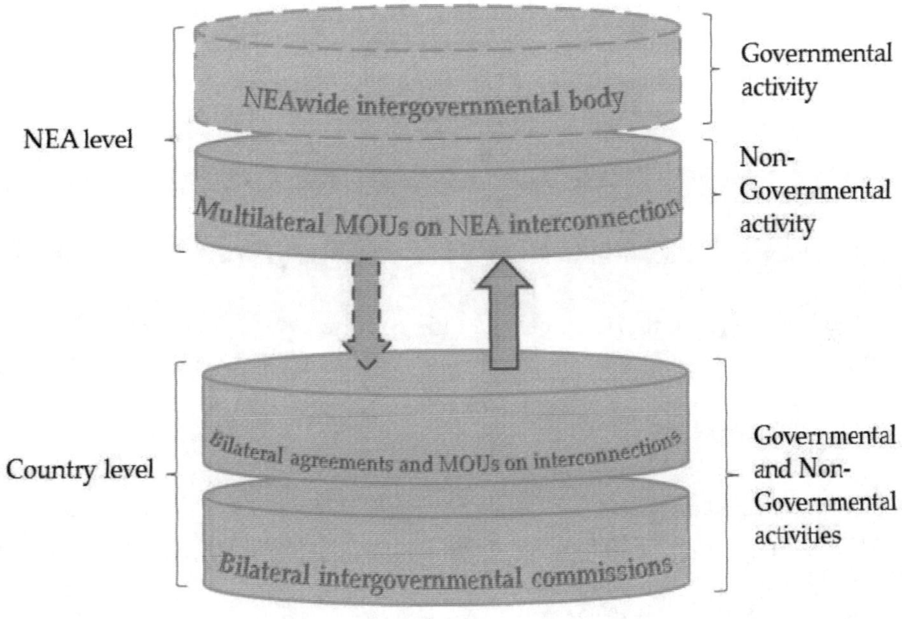

Fig. 5. Organizational/institutional multilevel hierarchical system for promote and development of NEA power interconnection.

Formation of this organizational and institutional infrastructure will help promote and develop mega-project of NEA-wide power system interconnection.

Conclusions

Electric power cooperation in NEA region is currently proceeding based mainly on bilateral intergovernmental and inter-company agreements.

Multilateral cooperation in the region is just emerging.

The world experience witnesses that electric power cooperation succeeds when special multilateral intergovernmental bodies promote interstate power cooperation.

Such a body needs to be established in NEA to make NEA-wide electric power cooperation more effective and intensive and to promote and develop power interconnection in the region.

References

1. L.S. Beliayev, S.V. Podkoval'nikov, V.A. Savel'yev and L.Yu. Chudinova, *Efficiency of Interstate Electric Ties* (2008)
2. F.T. Sparrow, H. Brian Bowen, Yu. Zuwei, *IEEE 2005 General Meeting,* 59 (San Francisco, 2005)
3. I. Syaiful, *Regional energy trade Workshop ADB* (Manila, 2014)
4. *Central Asia South Asia Electricity Transmission and Trade Project* (2014)
5. C. Brancucci Martinez-Anido, M.L. Vandenbergh, de Vries et al. Energy Policy, **61**, 207 (2013)
6. Al-Mohaisen Adman, S. Sud, *IEEE 2006 General Meeting,* 84 (Montreal, Canada, 2006)
7. *Regional Cooperation Strategy on Interconnected Power Networks in Indochina,* 47 (Tokyo, Japan, 2002)
8. *Energy Interconnections and Regional Integration in Latin America and the Caribbean* (2008)
9. *Power Interconnection in APEC Region. Current Status & Future Potential,* 82 (2000)
10. Zh. Liu, *Global Energy Interconnection* (2015)
11. L.S. Beliayev, S.V. Podkoval'nikov, V.A. Savel'yev and L.Yu. Chudinova, *ICEE 2015* (Hong Kong, 2015)
12. T. Otsuki, A.B.M. Isa, R.D. Samuelson, Energy Policy, **89**, 311 (2016)
13. D. Bogdanov, C. Breyer, Energy Conversion and Management, **112**, 176 (2016)
14. S.S. Shrestha, *Development of a Potential Regional Hydropower Plant in South Asia: A Prefeasibility Study* (2016)
15. B. Adeyemo, *South Asia Regional Workshop on Competitive Electricity Markets* (Colombo, SriLanka, 2014)
16. *The scheme and program for the expansion of the Unified Energy System of Russia for 2017-2023* (2017)
17. *Iner RAO EUS. Trading* (2017)

 https://doi.org/10.1051/e3sconf/20182701002

Institutional Issues of International Power Grid Development in Northeast Asia

Konstantin Korneev[1], *Darya* Maksakova[1], *Sergei* Popov[1,*]

[1]Melentiev Energy Systems Institute SB RAS, 664033 Irkutsk, Russia

Abstract. With regard to the initiative of "International power interconnections in northeast Asia" here examines the challenges which may hinder the establishment of a common energy system in region. The analysis is based on the APERC methodology for international energy cooperation in the APEC region, and on historical approach to the development of pricing mechanisms and policy in the European common market of electricity trade and system services. Mechanisms for the prevention and overcoming of the revealed problems of regional cooperation in the power sector are proposed.

1 Introduction

Prospects and problems of creating power interconnections are discussed in Northeast Asia region (NEA) for more than a quarter of a century. Currently there are international electric power connections between the four states: Russia and China, China and Mongolia, Russia and Mongolia, China and the Democratic People's Republic of Korea (DPRK). Table 1 shows consumption of electricity in 2015, the share of imports and exports for each NEA country. Despite of availability of the power interconnector infrastructure within region, Mongolia is the only state where international power trade shows significant role in national electricity balance.

Table 1. Import share in final electricity consumption of the Northeast Asia economies in 2015, percent.

Country	Electricity consumption	Import share	Export share
	TWh	*Percent*	*Percent*
China	4877	0.1 (0.1)[1]	0.3 (negl.)
Japan	949	-	-
Russia	726	0.9 (-)	2.5 (0.4)
ROC	495	-	-
DPRK	10	0.4 (0.4)	1.2 (1.2)
Mongolia	5.3	26.9 (26.9)	1.0 (1.0)

[1] In parenthesis – trade between NEA economies
Source: [1].

* Corresponding author: popovsp@isem.irk.ru

https://doi.org/10.1051/e3sconf/20182701002

The largest electricity flows within NEA are exports from Russia to China, and from China to Mongolia. The main electricity flows in 2009-2016 and the priority of the proposed projects are presented in Fig. 1.

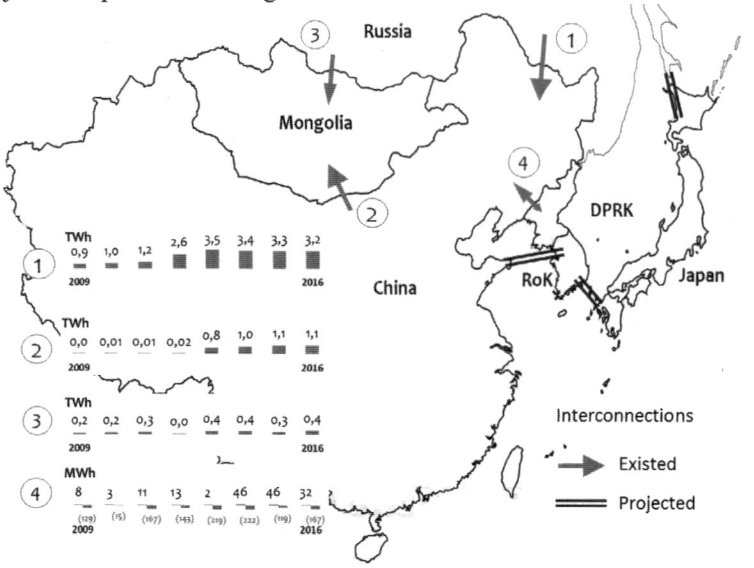

Fig. 1. International power interconnections in Northeast Asia in 2009-2016. Source: [2]

Electricity trade (also hereinafter referred as power trade to emphasize its power system particularities) within region is regulated by contracts. This institution requires systematic engagement of the bilateral institutions to the level of intergovernmental commission's competence, at least. Thus, development of new power interconnection project require legislative amendments, and gravity centre shifts from technical and economic issues to financial and political interstate relations.

Despite the fact that Japan and the Republic of Korea (ROK) do not participate in international power trade, the discussion for creation of a regional power network infrastructure is a novelty not only for them. The task to create a developed system of power interconnections change paradigm of energy supply for all countries in the region. For the energy business this means the need to develop (or establish in case of their absence) institutions for international power trade and relevant services within the region. States will need in adjustment of institutions shaping their energy policy, which in turn may affect social, economic and foreign policy frameworks of these states.

This paper deals with critical issues of creating international cooperation institutions imminent for the development of electricity infrastructure in northeast Asia in the long term. The methodological approach to the topic is based upon studies on the practice of multilateral energy cooperation in APEC [3, 4].

2 Purposes and characteristics of initiatives for develop power interconnections in Northeast Asia

It should be noted that there are a lot of projects related to the development of international energy infrastructure in Northeast Asia. All these initiatives differ in time frames, participants list, power infrastructure features, scope of the expected cooperation between national and sub-national power grids, *etc*. For the purpose of this study, the tasks that need to be addressed arose from an initiative of Northeast Asia Regional Power Interconnection and Cooperation Forum, initiated at the organisational conference held in October 2016 in Beijing [5].

The co-organisers of this international conference were China Electricity Council (CEC) jointly with Global Energy Interconnection Development and Cooperation Organisation (GEIDCO) and Energy Charters Secretariat (ECS). The conference participants included: representatives of governments, energy companies, industry and research organisations, as well as international organisations, notably – the Asian Infrastructure Investment Bank and the United Nations Economic and Social Commission for Asia and the Pacific. The goal of the new initiative was defined as the development of mutual understanding between the participants to facilitate the conclusion of agreements between regional governments to develop international power infrastructure within the Northeast Asia region.

An informal definition of this initiative's goal is the creation of united power system in the region, which will foster the most efficient use of generating and accumulating capacities in the participating countries to meet the needs of domestic electricity consumers. Thus the advantages of remote generation sources will be available to all consumers by means of transit through other member countries.

Steps to achieve this target will require the realisation of the following conditions:

- the rules, agreed by all parties of an interstate power grid, that regulate access to the transmission systems;
- peer participation of stakeholders in the development of transmission systems
- pricing mechanisms availability along the routes of power transmission between generators and consumers.

The latter is required to reflect the technical limitations set by for power generation and transmission system in order to meet the supply and demand balance. In addition, and most important for the coming decades, such pricing mechanisms (determined for short time frames) work effectively in power systems with a high rate of intermittent renewable energy generators.

2.1. The issue of multilateral cooperation

Connecting two hubs in different countries throughout transmission lines demonstrates the bilateral kind of power interconnections in a natural way. Multilateral aspects are possible only in the case of multiple connections between several countries, *i.e.* in case of the establishment of international power grid qua systematic progress of bilateral power interconnections. In this view, multilateral approach will allow to utilise the transit (and storage) potential of the power infrastructure to produce positive effects for all participants of international power grid. Numerous studies show the presence of synergy for such associations established in the region of East Asia. In particular, the salient economic effect appears for the case of increasing share of power, generated from intermittent renewables, and due to the operation mode of the hydropower to provide system services for maintaining power consumption. The most recent publications should be named are by Podkovalnikov et al. [6], Otsuki et al. [7], Otsuki [8].

Within the framework of the *Northeast Asia Regional Power Interconnection* International Energy Initiative (IEI *NEARPIC*), a major source of generation based on wind and solar energy is assumed to be arid and semi-arid areas of the Gobi and Western China, areas of the seas adjacent to Japan, the Korean Peninsula, and coastal provinces of China. The development of European united energy system, which brings together almost all the continent, except Russia and some countries of the former USSR, might be used as an analogy for IEI *NEARPIC*. The major issue for the governments at various levels in East Asia will be the determination of acceptable extent for national markets integration into a regional power grid in North-East Asia (Asian Super Grid).

Regardless of the dominant character of bilateral projects, the participation of several other organisations in the IEI *NEARPIC*, for example, international development banks and research organisations, can have a significant impact on accelerating the achievement of given initiative goals and increase the likelihood of transition to the stage of energy infrastructure's actual development. Such actors' contribution will be most needed for the development of power market regulation institutions: codes for buyers and sellers access to generation and transmission infrastructure; rules of capacity reservation and system regulation; pricing mechanisms; overall market regulation.

2.2 Contradictions between actors

The APERC study says there are three main groups of actors: businesses, governments (which includes international government-based organisations), civil society. For national and international business participation in the IEI *NEARPIC* provides additional opportunities in such areas like R&D, energy infrastructure construction, and operation. Gradual development of energy infrastructure will lead to international power interconnections, and furthermore – to a regional (international unified) power grid. It is important to note that mutual agreement and adaptation of national and international power market institutions are required for creating these opportunities and attracting businesses to such a project. These institutions include the mechanisms of financial, technical, environmental regulation that provide different types of grants, subsidies and guaranties for energy business. The process of reaching agreement on national and international markets institutions is comprehensive, iterative and bilateral. Besides, positive results for businesses may be non-obvious. The examples of the institutions that require agreement between national and international level are government guarantees, tax benefits, fid-in-tariffs, compulsory share of renewable energy in power generation, etc.

National and international government structures play a crucial part because they have an authority to influence on the rest of the IEA *NEARPIC* actors. National governments and international government-based organisations take all the issues related to achievement of positive balance for each actor's contribution and those social, economic, political gains he receives from international energy cooperation.

Imprudent government decisions can lead to contradictions between businesses and the other actors on national (or sub-national) level. The examples of such decisions are shown in the boxes "Economic consequences of power generation infrastructure reduction in import countries" and "Social and economic consequences of providing power maintenance services in export countries". In order to prevent negative consequences of the premature decisions, special mechanism should be implemented within the IEI *NEARPIC*. Such mechanism will assist any actor to examine different models of the international power markets institutions and assess they efficiency. These models can include different levels of informational openness and playground in the areas related to national energy infrastructure operation, regulation of infrastructure development, security criteria. The mechanism, based on different types of models with approved methodology, will be used to explain and validate reasoning along the policy making process, as well as provide opportunities for common language communication within stockholders [3, 4].

Economic consequences of power generation infrastructure reduction in import countries

Electricity import cuts off business activities in importing country, such as fuel production, transportation, conversion, power generation, from sources of profit. For example, through reduction of reserve capacities, the business associated to such generators should fade away. In this case cash flows and value added, related to the business, are withdrawn from the economy, and total volume of employment rate decreases. At the same time, the opportunity to import electricity can lead to discounted price for consumers that raise competitiveness of domestic production and the total level of economic welfare. Besides, the transfer of environment pollution sources to export countries decreases environmental impact of importers, thus transferring externalities and total cost of environmental abuse.

Social and economic expenses of providing power system services in export countries

The engagement in the system's capacity control raises pressure on the environment. For example, the use of hydro power plant reservoir as an accumulator within integrated power system requires mechanisms for adequate assessment of externalities cost. It might lead to contradictions between businesses and different levels of national authorities, and even rise to acute international issues of joint water systems regulation in exceptional cases. In order to compensate for negative economic and social effects, there should be robust mechanism for redistribution of the income derived at international power markets. This mechanism should ensure both overall positive effects for each affected stakeholder and sustained interest of all actors in cooperation at international level.

2.3 Reforming the national markets

Currently, there are no countries with a truly competitive power market in the East Asia region. The reform of power industry in Japan is still in a stage of testing recently introduced competition rules. While Japan and the Republic of Korea have achieved significant success in introducing market institutions in the electricity sector, they requires time to adapt. The indicator of a mature competitive market is a relatively low threshold for entry of foreign business, such as power generator, electricity trader or network services provider. Although Russia is one of the first and the most advanced economies in introducing competitive mechanisms within power industry in the Northeast Asia, the Russian power wholesale market is not introduced at the territories, related to the IEI *NEARPIC*. The one exception is the Baikal region, representing the most eastern part of the power and system services market in Russia. Yet creation of national competitive and liberalised markets in China and Mongolia are under discussion, there is no real advancement so far. Not participating in the IEI *NEARPIC* process, the DPRK appears to be far from discussion on the rules of cross-border power connections. The existing bilateral trade relations between Russia, China and Mongolia are based on long-term contracts, where the price mechanisms do not reflect the balance on fast evolving national markets for power and network services.

Great efforts in the creation and development of international power and system services trading institutions will be required from all the participants in order to integrate several countries and transform bilateral power interconnections into the international power grid of the Northeast Asia region. Let us assume there are submarine cables between the Shandong province in China and the city of Incheon in the Republic of Korea, between the Republic of Korea and Japan through the Strait of Tsushima. The grid infrastructure is technically able to transfer the power, generated by wind turbines at the south of Mongolia to the Western Prefectures of Japan through China and The Republic of Korea. It is obvious that there should be institutions to regulate the procedure of signing standardised sales/purchase contracts for a unit of capacity between participants from at least two participating countries with minimal transaction costs. Capacity purchase is a deal where the seller guarantees capacity at certain voltage level, by certain discretion of capacity for certain discretion of time, e.g. "*5MW at 500kV voltage for 15 minutes from 11:30 to 11:45 at X node of the X-Z transmission line*". Currently the following binding instruments for the implementation of this transaction are missing:

1. mechanisms for setting a "day ahead" price of a unit of power capacity (UPC) at the Mongolia's wholesale market
2. tariff for the transmission of UPC by the national transmission lines of Mongolia (hereinafter power transmission takes losses into account)
3. custom tariffs on the UPC exported from Mongolia
4. custom tariffs on the UPC imported to China
5. charges, related to injection of UPC into the Chinese power grid
6. transit tariff for the transmission of UPC by the national transmission lines in China from the point of import (from Mongolia) to the point of export (to the Republic of Korea)
7. similar to points 3-6 for UPC imported from the Republic of Korea to the power grid of Japan
8. mechanisms for setting a "day ahead" price of UPC on the wholesale market of Japan.
9. real-time wholesale market to balance consumption and generation for all national power grids in the countries involved
10. post-factum clearing mechanisms to settle forward and balancing deals

In case of significant purchases and sales of renewable power, or in case of utilising large energy accumulators in the system (e.g., pumped storage and/or hydropower plant), or in case of providing critical service to maintaining power quality within the system, the institutions to ensure such transactions are necessary.

The examples of essentiality for the regulation of system services are the mechanisms for incorporating the "externalities cost" into the price of electricity supplied by hydropower plants. There are different operational modes for hydro plants. The first mode is standard, corresponding to the power generation from the controlled source of renewable energy. The second mode refers to the participation in the process of energy accumulation within power grid. An issue of seeking "a fair price" is crucial for creating common rules to determine commercial and economic efficiency for each actor.

The differences in national regulatory framework for power industry will have the most significant impact on the development of pricing mechanisms, which in turn represent a key issue for international power interconnections. Harmonisation of the market regulation institutions for all economies involved in the IEI *NEARPIC* will provide a key to the very success of the international cooperation. Building a joint system for regulation of international trade of power and power system services requires enormous amount of work to bring the legislation of all participating countries to common denominator. Since international power trade has significant impact on social and economic development of the participating countries, it is necessary to ensure working mechanisms of energy strategies coordination for each country, among other types of the common power market institutions. The latter circumstance requires special attention from all of the IEI *NEARPIC* participants, since it may result in a radical revision of the current energy security paradigm.

2.4. The issue of insufficient awareness of the participants

Currently there is no coherent plan to create a system of international power interconnections in Northeast Asia. The lack of tools that would allow a wide range of participants to assess the effectiveness of their participation in IEI *NEARPIC* based on shared vision can complicate and slow down the process of achieving the goals of this international initiative. Moreover, such a major infrastructure project like the construction of the interstate power connections in Northeast Asia has significant multiplicative effects.

Namely, there is an impact on the development of important industries, like mechanical engineering, metallurgy, chemistry, construction. Using the definition of goals set for the IEI *NEARPIC*, this international initiative is expected to solve following tasks in the informational context:

a) to create a platform for constructive discussions and stimulate the search for alternative approaches for issues arising from the international institutions' implementation process, and thus ensure the effective development of interstate power interconnections in Northeast Asia;

b) to spread knowledge on the nature and benefits of cooperation between participating countries, adapted to various segments of the audience.

Creation and development of a mechanism that allows solving these issues on a basis of mutual trust can form bold foundation for this initiative. As it follows from empirical studies described in the second APERC report on international energy initiatives [4], the trust between actors arise on awareness, underpinned by a commitment to achieve common goals (with their adequacy to the circumstances external to the initiative's framework). That postulates the basic prerequisite for the successful implementation of any international energy cooperation.

One kind of inter-actors communication mechanisms can be designed upon modern network resource, with graphic content and gaming component, in order to enhance awareness and intensive considerations of new power interconnection projects. In the form of business game it will provide an opportunity for each participant to consider his own project evaluations at different levels of aggregation relying on proven methodology – from subcontractors to local generators, to regional (or national) government. Another purpose for such a mechanism will be a background preparation for discussions at the working group level.

2.5 The issue of technological scope

To a certain extent the goal, established within the IEI *NEARPIC*, is due to the political requirement to reduce nation's dependence on fossil fuels. One of the most viable approaches to deal with this issue is to accelerate development for renewable power generation. Taking into account technological progress, that means a growth of wind and solar energy share in total electricity production. Additionally, due to intermittent nature of wind and solar energy, the needs to ensure sufficient capacity and topography for transmission lines, and electricity accumulation capacities within power grid, should be taken into account.

In the case of bilateral power connections establishment, even more in case of the Northeast Asia international power grid development, there will be a need for an organisation that holds confidence of all participants to provide guaranties and reduce risks. In such a case, the national energy management systems should be harmonised at the institutional level of multilateral international cooperation.

While at the initial stage the IEI *NEARPIC* is seeking for its ultimate goal and patterns to reach them, it seems appropriate to implement the experience of similar initiatives being undertaken in this region. Established in 1996, the IEI *NAGPF* set its goal as "*to develop the concept of creating a network of international gas trunk pipelines in the Northeast Asia*". Since gas trunk pipelines relates to the technologies of continuous energy transportation, it can be considered as some analogue for power systems. From the very beginning the IEI *NAGPF* principally distanced itself from the liquefied natural gas (LNG) infrastructure, which is based on discrete transportation technologies. The opportunities were missed to consider different but complementary segments of a single market. Two decades later, we have to admit that this approach had significantly reduce importance of the whole initiative for all participating countries to deal with issues of energy security and domestic market reforms.

Insufficient accounting of technological features could lead to incomplete evaluation of the market segments. For example, the expected commercialisation of the power-to-gas, power-to-liquid technologies will create new markets. Such technologies will became competitive to established technologies for power transmission through wires in terms of energy security, reliability and economic efficiency. Consequences of such technological factors impact inevitably will lead to important transformation of the energy markets institutional structure.

The discrete alternatives for electricity transmission have several advantages, related to security and robustness of energy infrastructure:

1. It allows eliminate the gap between generation and consumption due to accumulating abilities of the discreet energy transportation subsystem.
2. It allows escape the monopsony situation by exploiting ability to diversify transition routes in absence of fixed routes infrastructure.
3. It has no extremely high requirements for robustness of long-distance transportation system to keep power supply security.
4. Domestic consumers of energy services can benefit from enhanced competition between different types of energy carriers at national energy markets and from increased level of energy supply security in general.

The existence of cost-effective alternative technical systems to provide energy services (mechanical movement, lighting, air conditioning, *etc.*) does not mean that the development of power interconnections in the Northeast Asia does not require complex multilateral agreements as precondition. However, considering alternatives can help in evaluation of their (multilateral agreements) structure, scale and sequence of evolution [9].

2.6 The DPRK Issue

Weak representation, or lack of participation of the DPRK in the *NEARPIC* initiative reduces the effectiveness of the entire project, since it excludes the use of its hydropower facilities as regulators (and/or accumulators), and the territory for the construction of power interconnections. Yet, provided the positive economic experience from power interconnections is obtained, and the openness of the IEI *NEARPIC* for the accession of other participants is gained, both would establish a powerful factor to involve the DPRK to the regional energy cooperation activity. In addition to positive impact on the socio-economic development of participating countries, the involvement of the DPRK into the process of negotiating and discussing the institutions of power interconnections development undoubtedly will lead to a much healthier political climate within the region.

The United Nations Economic and Social Commission for Asia and the Pacific (ESCAP) is representative of the highest level of international intergovernmental actors in IEI *NEARPIC*. The role of this organisation by definition is to act as a mediator between governments to assist in solving common social and economic issues, and it is especially important due to the significance of political factor in the East Asia region. Currently, ESCAP is one of the most influential and trusting actor within the IEI *NEARPIC* process, and by its status has the opportunity to engage the DPRK into the process. The format of the initiative allows actors to discuss technical, economic, social and financial regulation emerging during the implementation of joint projects, and assess the participants' economic and social effectiveness, while leaving political factors behind the scene, or reducing them to technical parameters.

When considering the issue of the DPRK involvement, one should take into account the negative experience of similar initiative on energy cooperation at the Korean Peninsula, which was doomed to failure from the very beginning because of an incorrect goal setting (see "The KEDO Initiative" box).

KEDO Initiative

The negative environmental assessment was the official reason for termination of the project to build a nuclear power plant in the DPRK – the project, proposed by the United States, the ROK and Japan in March 1995. The KEDO project (Korean Energy Development Organisation) initiative was offered in exchange to the DPRK's refusal from constructing a nuclear power plant, based either on the Soviet, or domestic technologies. The main argument for the negative environmental assessment was aroused from discrepancy between the capacity of proposed power plant (two 1GW reactors) and existing status of the DPRK power grid. In this country a single generator capacities do not exceed 200-300 MW, and the country's backbone transmission lines does not guarantee safe operation of nuclear reactors due to overall grid inefficiency.

Experts had foreseen such results long before, after abandonment of the first nuclear power plant project north to DMZ. On a wave of consequences after USSR dissolution, project for several 400 MW reactors of the USSR design had been substituted by another project on the same site. That new project implements US-based technologies for two 1000 MW units, which actually could not fit to the DPRK power system. Petroleum deliveries to the DPRK dried up in 2003, following negative conclusion of the project's environmental impact assessment. Overall, the process lead to the second withdrew of the DPRK from the Nuclear Non-Proliferation Treaty. The first time the DPRK leave the Nuclear Non-Proliferation Treaty in year 1993, after failure of the comprehensive safeguards regime implementation under agreement with the IAEA (came into force in April 1992).

The Directorate of the KEDO program on May 31, 2006 decided to stop any activity for the reactor construction. As a consequence the existed minimal level of confidence in both nuclear non-proliferation and energy cooperation between the DPRK and the KEDO member countries was eliminated [10].

Implications

The strengthening of integration processes in the Northeast Asia region and the concept of a single electricity market development are behind the elaboration of options to create regional power grid. The Northeast Asia economies have favourable economic and geographical conditions, appropriate technologies, but lack of experience in energy cooperation is a significant barrier for such a project. Dramatic transformation for power market institutions, as well as rethinking of principles of the national energy policy will be required for each participant in such a complex integration project.

International energy initiatives can become an effective mechanism for energy cooperation within the region. The project of such an initiative - *NEARPIC* - is intended to provide the platform to facilitate the development of power interconnections in the region through the preparation of favourable framework for multilateral cooperation. The main task at the current stage is to organise the search for proper international cooperation mechanisms, which will ensure the effectiveness of such cooperation for each actor. The need to create a methodology for assessing the effectiveness, and tools for practical use derives from the main task. On the basis of such mechanisms, it will be possible to carry out comprehensive assessments of the Northeast Asia power industries synergy, seek a balance of interests among various actor groups within multilateral framework, like energy business, governments and intergovernmental organisations, and civil society.

The authors consider major factors impeding effective Northeast Asian countries cooperation on the path to unified regional electric power infrastructure:

- Contradicting interests of participants, qualitative heterogeneity and quantitative incommensurability of benefits and costs for each actor.
- Unequal status of the national power market institutions within Northeast Asia countries is leading to inconsistency of national financial, technical and environmental regulation.
- Complicated political stage within the Asia-Pacific region (and Northeast Asia in particular), supplemented by weak mutual understanding in long-term energy policy.
- Lack of coordinated information support for international projects in power industry, leading to a lack of understanding of already announced initiatives for participants, non-commercial organisations and population.

Nevertheless, successful overcoming of the above-described problems is quite possible, which is proved by the example of power market development in Europe. A clear understanding by all participants of issues and priorities for the creation of Northeast Asian joint power grid will positively affect the development of such complex forms of cooperation as power interconnections within the region.

References

1. http://www.iea.org/statistics/ (Accessed 25.09.2017)
2. https://comtrade.un.org/data (Accessed 27.09.2017)
3. S. Popov, D. Fedor, *Understanding international energy initiatives in the APEC region: scope and elements* (APERC, Tokyo, 2007)
4. S. Popov, D. Fedor, *Understanding international energy initiatives in the APEC region* (APERC, Tokyo, 2008)
5. S. Podkovalnikov, V. Savelyev, L. Chudinova. Bulletin of the Russian Academy of Sciences: Energy, 2015, vol. 5, pp. 16-32
6. T. Otsuki. A. Mohd Isa, R. Samuelson. EP, vol. 89 (2016), pp. 311-329
7. T. Otsuki. RE, vol. 108 (2017), pp. 321-335
8. http://www.unescap.org/sites/default/files/Concept%20Note%20for%20the%20Joint%20Conference%20on%20Northeast%20Asia%20Regional%20Power%20Interconnection%202017.pdf (Accessed 15.09.2017)
9. S. Popov. Energy Policy 3, 2013 (in Russian)
10. http://www.kedo.org/au_history.asp (Accessed 19.09.2017)

Analysis of the prospective energy interconnections in Northeast Asia and development of the data portal

Andrey Churkin[1,*], *Janusz* Bialek [1]

[1] Skolkovo Institute of Science and Technology (Skoltech), Center for Energy Systems, 143026 Moscow, Russia

Abstract. Development of power interconnections in Northeast Asia becomes not only engineering but also a political issue. More research institutes are involved in the Asian Super Grid initiative discussion, as well as more politicians mention power interconnection opportunities. UNESCAP started providing a platform for intragovernmental discussion of the issue. However, there are still a lack of comprehensive modern research of the Asian Super Grid. Moreover, there is no unified data base and no unified power routes concept. Therefore, this article discusses a tool for optimal power routes selection and suggest a concept of the unified data portal.

Introduction

The idea of creation Asian Super Grid (also mentioned as Asian Super Ring) as well as initiatives of direct power export from Russia to the Republic of Korea and Japan are not new. The issue of Asian Super Grid creation was mentioned by RAO UES (Unified Energy System of Russia) in 1998-2000. Options of direct power export to Japan were also considered. In 2003 RAO UES has signed memorandum of understanding with Japanese corporation Sumitomo. The necessity of Asian Super Grid development was mentioned several times by president Vladimir Putin at Asia-Pacific Economic Cooperation (APEC) forum and economic forum in Vladivostok. The president of Russian Federation expressed his support of Russian and foreign initiatives of energy integration and suggested to create a working group.

In recent 15 years several researches and analysis were performed on the Asian Super Grid. Most of them are listed in [1]. There also should be mentioned Skoltech research on power export options to the Republic of Korea and Japan, ESI SB RAS research on Asian Super Grid effectiveness and Gobitec project analysis [2], and recent researches of Renewable Energy Institute, Japan. At the Joint Conference on Northeast Asia Regional Power Interconnection 2017 it was shown that many other institutes and companies perform their own studies and analysis on power interconnections in Asia: Korea Electric Power Corporation KEPCO, Inner Mongolia Power Group, NovaTerra LLC, China Electric Power

* Corresponding author: andrey.churkin@skolkovotech.ru

Planning & Engineering Institute, Korea Electrotechnology Research Institute (KERI), and others.

However, there is no unified scheme (power interconnection routes) in researches mentioned. A difficulty exists in comparing the researches and initiatives due to lack of open information. This article suggests a tool for optimal power interconnection routes selection and data portal concept that contains information about all existing researches, analyses and initiatives of power interconnection. Development of such portal will aggregate information and present it form accessible even to non-technical people, promoting the Asian Super Grid project.

1 Analysis and simulation of existing power systems

Even though there are few considerable interstate power interconnections in Northeast Asia (only Russia-China power export via 500 kV line is significant), each solitary power system has already been formed, has its own topology, features and issues. High level researches performed suggest power interconnections arbitrarily in order to estimate approximate cost of the project. There is no analysis of optimal power interconnection routes and possible regimes. Such analysis requires accurate simulation of existing power systems.

We suggest model development of each Northeast Asian power system taking into account HV and UHV power lines. This kind of modelling has been successfully performed by professor J. Bialek in case of European Union [3]. The direct current (DC) model was composed on the basis of publicly available information. Then, simulated power flows were compared with actual power flows on a certain date. Average accuracy was over 90%. The model was later used by consulting agencies and institutes, particularly for renewable power sources integration effect analysis.

Skoltech started simulating North and Northeast power systems of China (fig. 1) because these regions would be the main chain of the Asian Super Grid, interconnecting Russia, Mongolia and Korea. The model is being composed on the basis of available information on power systems topology, demand and generation. At this stage the model contains 170 buses, 360 lines, 110 generators and 150 loads.

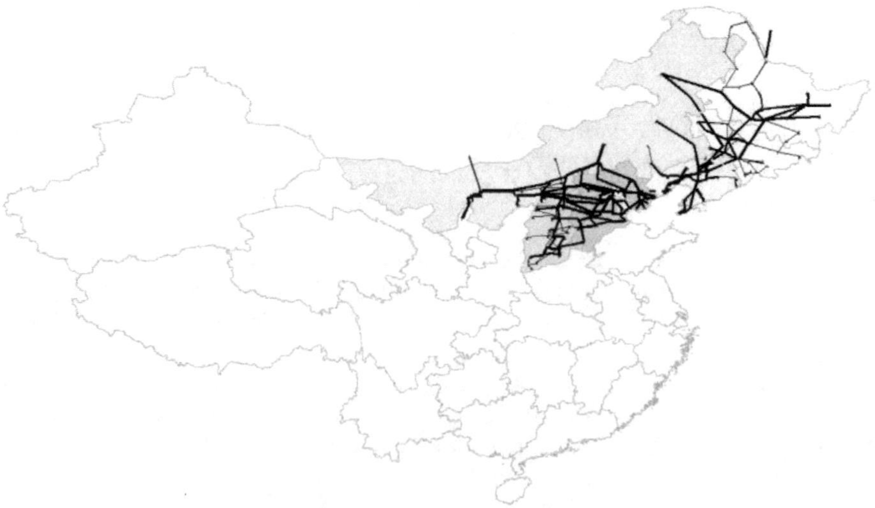

Fig. 1. North and Northeast Chinese power systems model

However, accuracy of the model is unsatisfactory. It can be stated that only directions of power flows do match to State Grid Corporation of China reports. Unsatisfactory results can be explained by:

• lack of publicly available information (most of the reports are not available or written only in Chinese),

• rapidly developing power system (dozens of HV and UHV power lines are planned or already under construction in China that makes available power system schemes outdated; each new powerful line changes power flow distribution significantly),

• features of the Chinese economy (highly industrialized country; huge amount of powerful factories; missing/neglecting one of the powerful loads decrease accuracy of the model).

Therefore, simulation of Northeast Asia power systems is a complex task that should be performed jointly by international institutes and companies. The model will enable to select optimal power interconnections, analyse power regimes, detect power systems bottlenecks, analyse effect of renewable generation injection, etc. Thus, the unique powerful tool can be developed that will bring Asian Super Grid researches to a new level.

2 Data portal concept

As was mentioned above, discussions and studies of power interconnections in Northeast Asia are performed for more than 15 years. However, there is still no unified concept of power interconnections. There is a lack of publicly available information. Results of the researches performed are not easy to find and compare. Therefore, it is hard to convince common people in Asian Super Grid project necessity and effectiveness.

Together with UN ESCAP Energy Division we developed a concept of Energy Connectivity data portal, containing all relevant up-to-date information about power interconnection initiatives. The concept can become the new section of already existing portal http://www.asiapacificenergy.org/ that aggregates reliable statistics and information about energy development in Asia and the Pacific regions. The new section will contain subsections that show power systems development, forecasts and initiatives of power interconnection. Concept of these subsections are presented below.

2.1 Power systems development map

The concept of power systems development map implies presentation of each power system as a circle with square proportional to total annual electricity production and demand (sum of power generation TWh and power demand TWh). Moreover, each circle can be divided as a pie chart in order to show the ratio of generation and consumption. The next pie chart around the circle shows ratio of primal energy resources import/export/production. The bar chart near the circle shows comparison of installed capacity and peak demand in a certain power system. Finally, the money bar shows an average electricity price in a system.

Once user opens the power systems development map, all necessary explanations should be provided. An enlarged example of power system charts and bars is presented in figure 2. The concept of power systems development map in case of Northeast Asia in shown in figure 3. The charts and bars represent data from 2013. Thus, power systems development map enables even non-technical users to compare different power systems, examine electricity prices and dependence on energy import. Moreover, the map will contain timeline that allows to examine power systems retrospective and forecast for future development.

 https://doi.org/10.1051/e3sconf/20182701003

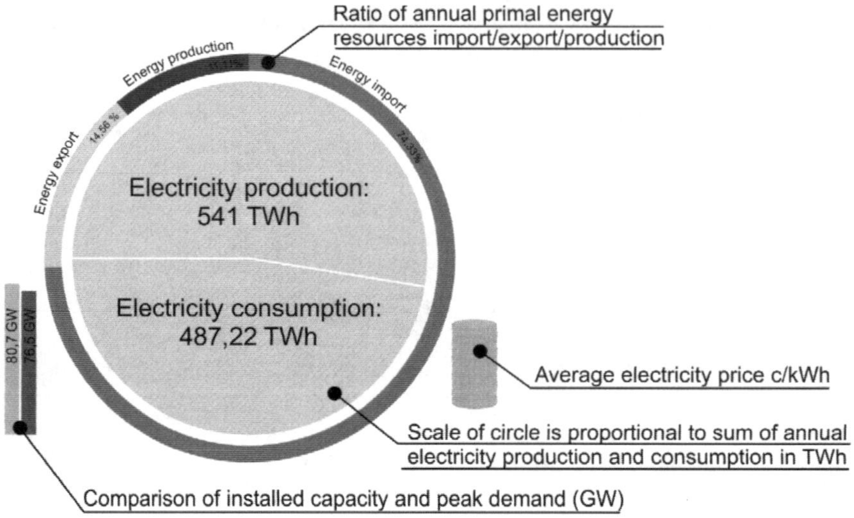

Fig. 2. Description of power system charts and bars

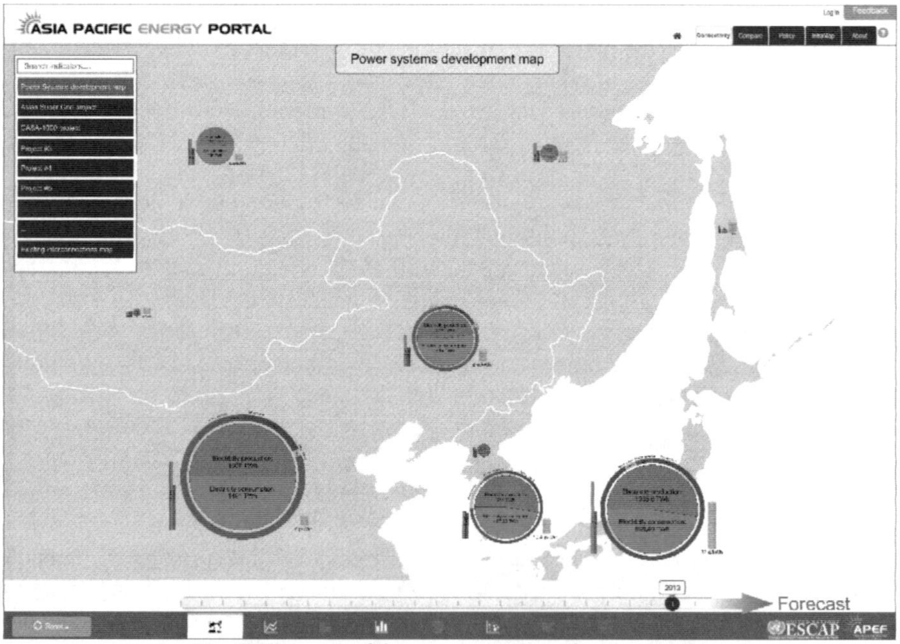

Fig. 3. Concept of the power systems development map

2.2 Project mapping

Despite power systems development map Energy Connectivity section will contain the list of existing projects of power interconnection in Asia. Each project involves at least several countries. Once user selects a certain project an interactive map opens allowing user should to examine initiatives of each country and results of researches on current integration project. There should be an option to switch on any layer of a project or all layers together.

There are numerous parameters that can be shown for a country in order to promote energy integration project. Moreover, these parameters are not the same for different countries. Some countries do not disclose information that their neighbors do. Most of the parameters can be presented in form of diagrams. Scheme of suggested interconnections can be shown on a map with explanations provided.

Analysis of power systems have been performed for Northeast Asia countries in order demonstrate necessity of the Asian Super Grid project, the most ambitious one. Information is shown county by county, as it would be in the Energy Connectivity portal.

2.2.1 The Republic of Korea

Considering the Republic of Korea high dependence on energy import should be mentioned (figure 4a). Through the last decade Korean rate of dependence in final energy consumption exceeded 95% [4]. The main energy import products are crude oil, coal, natural gas and oil products. Moreover, Korea has the highest rate of electricity consumption per capita Asia (figure 4b). Along with high rate of import dependence, this poses serious issues of energy security. In addition, electricity price is rather high [5] (figure 5b). Under these circumstances, Korea turns its attention towards renewable energy. In preliminary version of the 8th basic plan for long-term Electricity Supply and Demand it is suggested to phase out nuclear power by renewables whose share can reach 20% of total generation by 2030 (figure 5a).

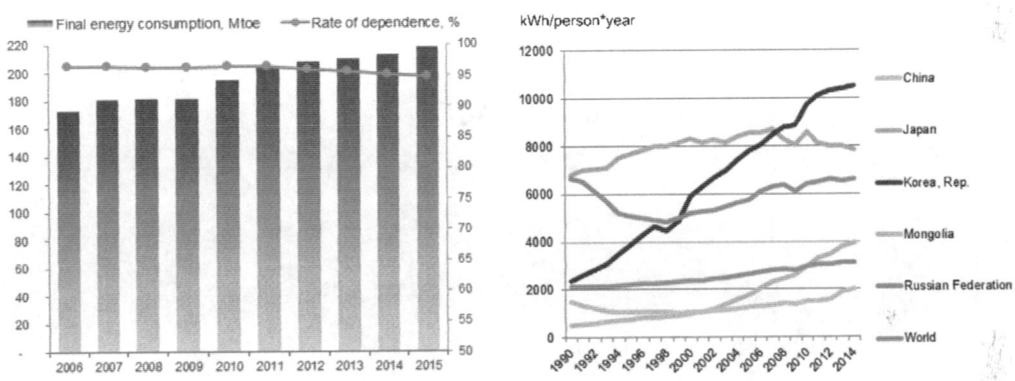

(a) (b)

Fig. 4. Korean energy consumption and dependence on import (a), electricity consumption per capita (b)

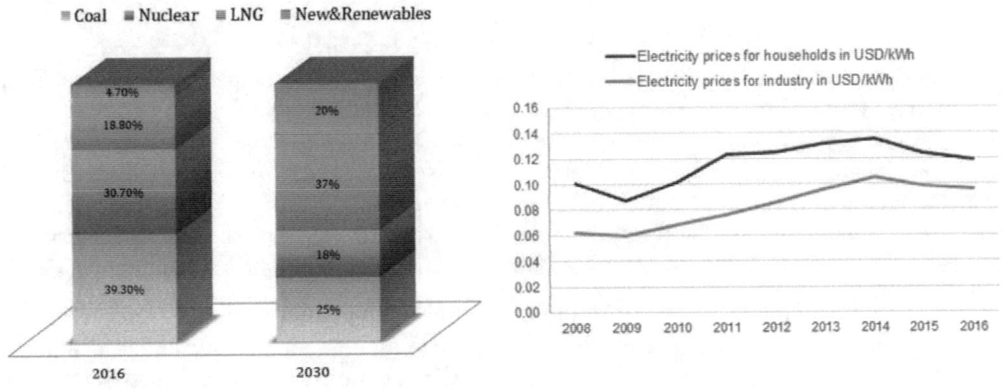

(a) (b)

Fig. 5. Korean energy policy shift (a), electricity prices (b)

 https://doi.org/10.1051/e3sconf/20182701003

Such significant increase in renewable generation share will influence power system operation and stability. The obvious measure of stability enhancement is interconnection with neighboring power systems. Therefore, possible options of interconnection have been studied (figure 6). The main routes are presented in the figure 6. However, most of the studies considered power lines construction through the Korean peninsula that cannot be achieved due to political issues. Recent studies suggested undersea cables construction to China and Japan. Preliminary assessment performed by KEPCO showed that 2 GW undersea interconnection China-Korea-Japan would cost 6 bln. $.

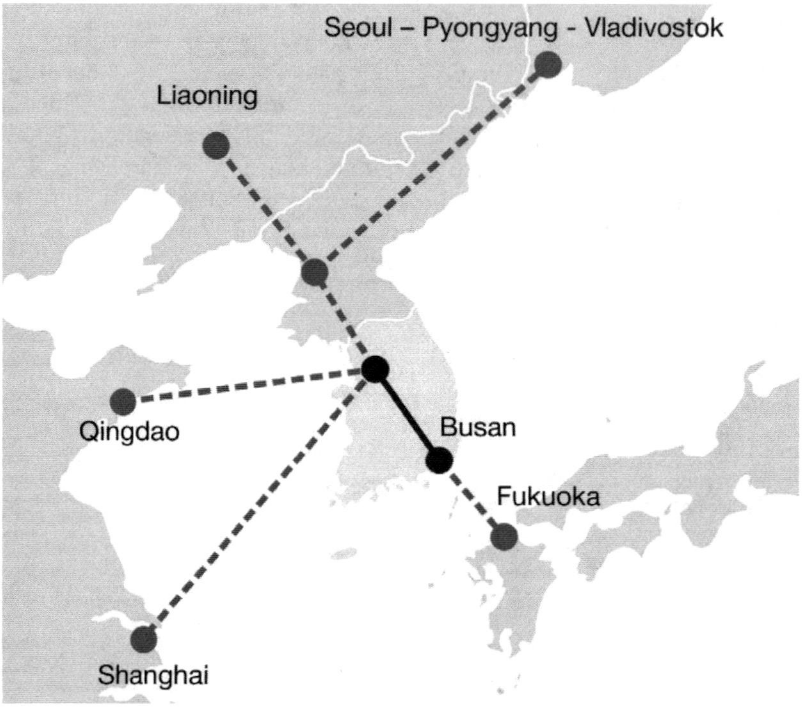

Fig. 6. Possible routes of power interconnection with the Republic of Korea

2.2.2 Japan

Japan is in similar conditions as the Republic of Korea being one of the most energy dependent countries in the world [6] (figure 7a). Electricity price in Japan is even higher than in Korea [5] (figure 7b). Therefore, Japan is also interested in renewable energy development that is presented in Long-term Energy Supply and Demand Outlook 2015 [7] (figure 8). It seen that Japan takes significant conservation measures in order to restrict power demand growth. Moreover, it is anticipated that annual power demand growth would start declining by 2030 and would be -1.2%. Regarding power source mix, Japan also forecasts significant increase of renewable energy up to 19 – 20 % of total generation by 2030.

It is natural that power interconnection with neighboring countries have been considered. Possible routes of interconnections with Korea, Japan and Russia are presented in figure 9. Construction of the power interconnection would alleviate Japan primal energy resource dependency and positively influence on electricity price. However, each interconnection project needs precise technical and economic assessment.

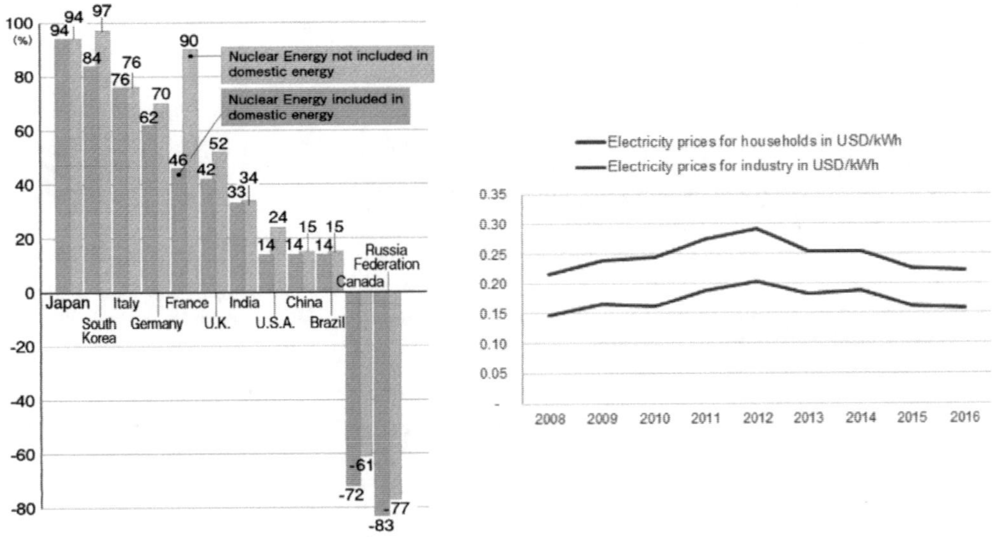

(a) (b)

Fig. 7. Japan dependence on imported energy sources (a), electricity prices (b)

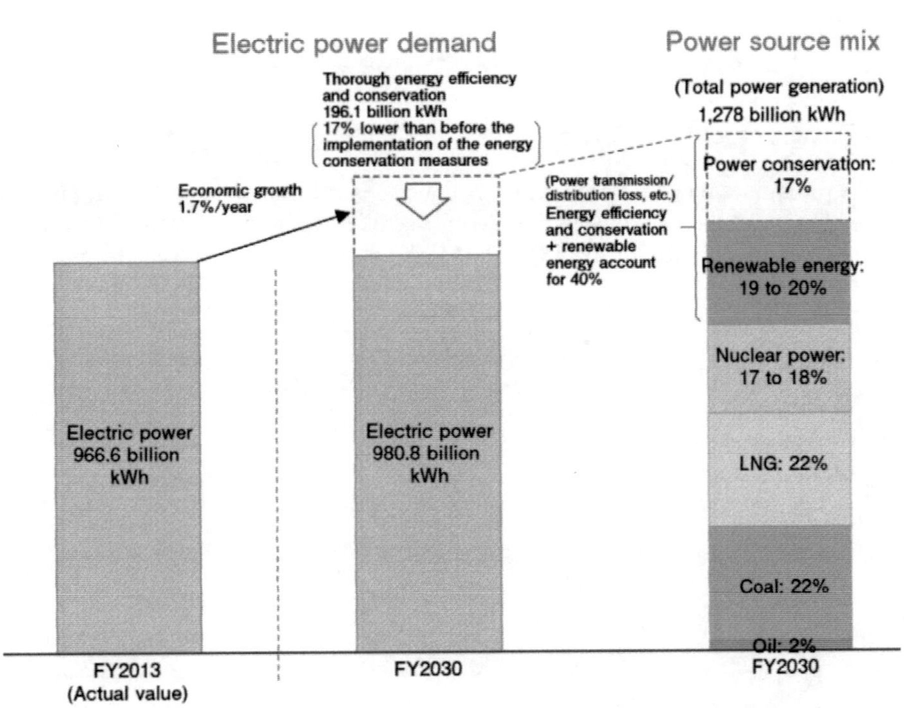

Fig. 8. Japan Long-term Energy Supply and Demand Outlook 2015

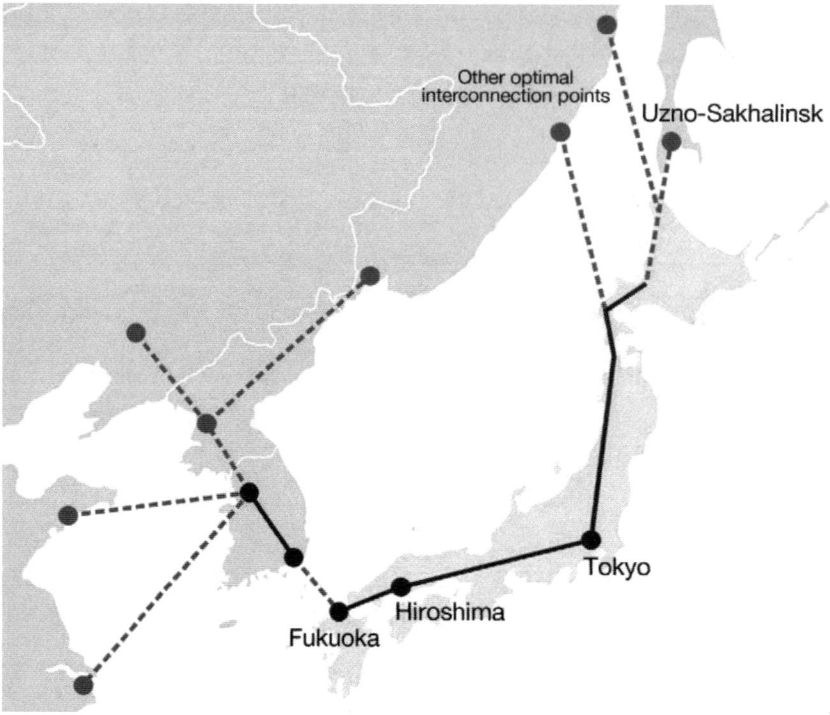

Fig. 9. Possible routes of power interconnection with Japan

2.2.3 People's Republic of China

It is hard to imagine interstate power interconnections in Asia without China whose power system development pace is outstanding. Total installed capacity growth of Chinese power system is presented in figure 10 [8]. The 13th Five-Year-Plan forecasts annual energy consumption growth by 2.5%, electricity consumption growth by 3.6-4.8% and GDP growth by 6.5-7%.

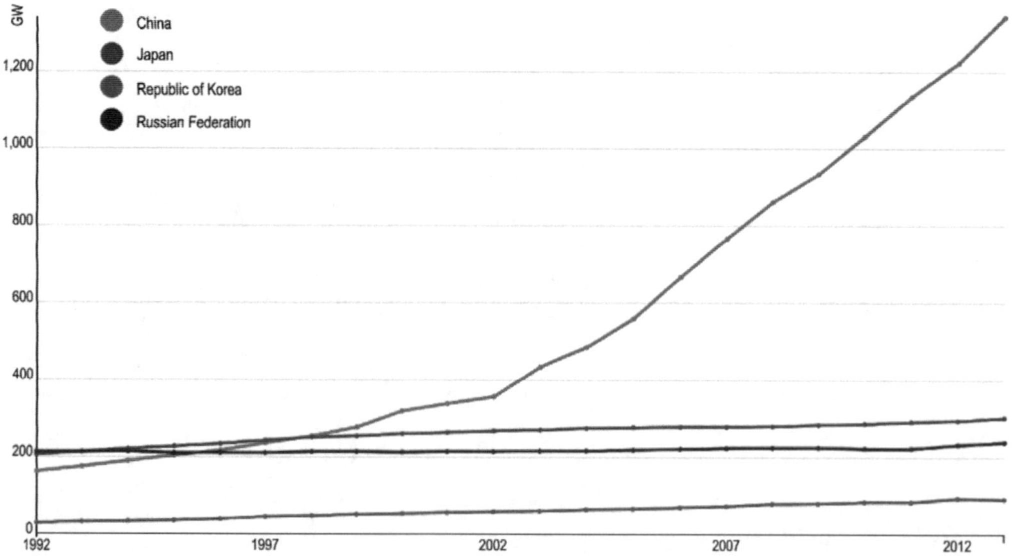

Fig. 10. Total installed power plant capacity statistics

China possesses high-developed energy infrastructure with more than 640 000 km of HV and UHV power lines. There also are existing interconnections with Mongolia and Russia. The infrastructure mentions creates an advantageous conditions for interstate power interconnection. Possible routes of power interconnection with China are presented in figure 11. It should be mentioned that China also develops renewable energy: the 13th Five-Year-Plan implies huge investment in hydro power and renewable energy up to 370 bln.$ by 2020.

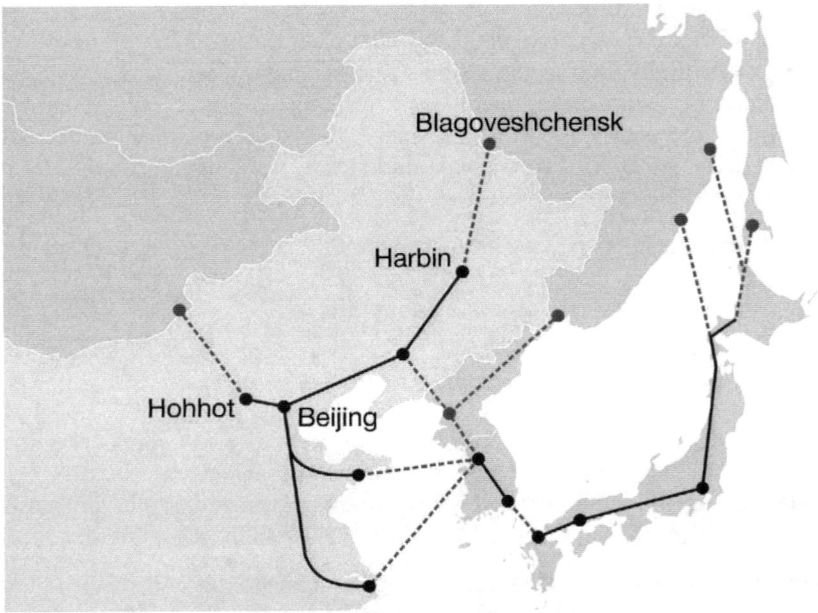

Fig. 11. Possible routes of power interconnection with China

2.2.4 Mongolia

Mongolian power system is small compared to neighboring systems. It possess 1.2 GW of installed capacity and is divided into 5 separate power systems. There is power import both from China and Russia. Nevertheless, Mongolia is interested in power interconnection due to its geographic position and renewable energy potential: Gobitec initiative implies renewable power sources creation in the Gobi Desert with installed capacity exceeding 100 GW.

Possible power interconnection would connect China and Russian Ulan Ude through Ulaanbaatar and Gobi Desert. This interconnection would enhance Mongolian power system stability and pave a way for renewable energy export.

2.2.5 The Russian Federation

Three Russian power systems are usually considered during Northeast Asia power interconnections discussion: Siberian power system, power system of the Far East and Sakhalin power system. The main point for participation in the integration projects is Russian huge hydro potential, especially in Siberia, and relatively low electricity price. Even nowadays, there is significant amount of power reserve and unrealized reserve in Siberia and the Far East: 7.1 GW and 4.1 GW consequently [9].

Installed capacity of Siberian power system and the Far East power system in 2017 reaches 51,97 GW and 9,19 GW respectively. Hydro power share reaches 49% in Siberian power system and 36% in the Far East power system. There are dozens of power plants construction projects and initiatives by “Rushydro”, “InterRAO”, “Eurosibenergo” that can increase installed capacity of Siberian power system up to 67,3 GW and capacity of the Far East power system up to 16,1 GW by 2035 (according to joint Skoltech – ESI SB RAS research, 2015). Installed capacity of Sakhalin power system can increase from 850 MW up to 1290 MW by 2035.

There is significant infrastructure in Sibria and the Far East (figure 12), as well as existing interconnections between Russia and China, Russia and Mongolia. The Far East power system is connected to Northeast China by 3 power lines: 110 kV, 220 kV and 500 kV that enables to export more than 3 bln. kWh per year. Power export from Siberia to Mongolia is less significant: there is 220 kV line with maximum transmission capacity 245 MW.

As an energy exporting country Russia has an obvious interest in Asian energy integration. Moreover, Russian geographical position enables to interconnect most of the Asian Northeast countries. Existing infrastructure and excessive capacity makes it real to discuss power export options in the foreseeable future.

2.2.6 Asian Super Grid

After analysis of each country’s power system and initiatives of interconnection, it is possible to present the entire scheme of the project (figure 12) that encompasses the Republic of Korea, Japan, China, Mongolia, Russian Siberia, the Far East and Sakhalin. Kazakhstan can also be included due to its proximity and Russia-Kazakhstan existing interconnection.

Subsections with existing interstate power interconnections and results of researches performed should be presented. Figure 13 shows economic benefits estimated by joint Skoltech – ESI SB RAS research.

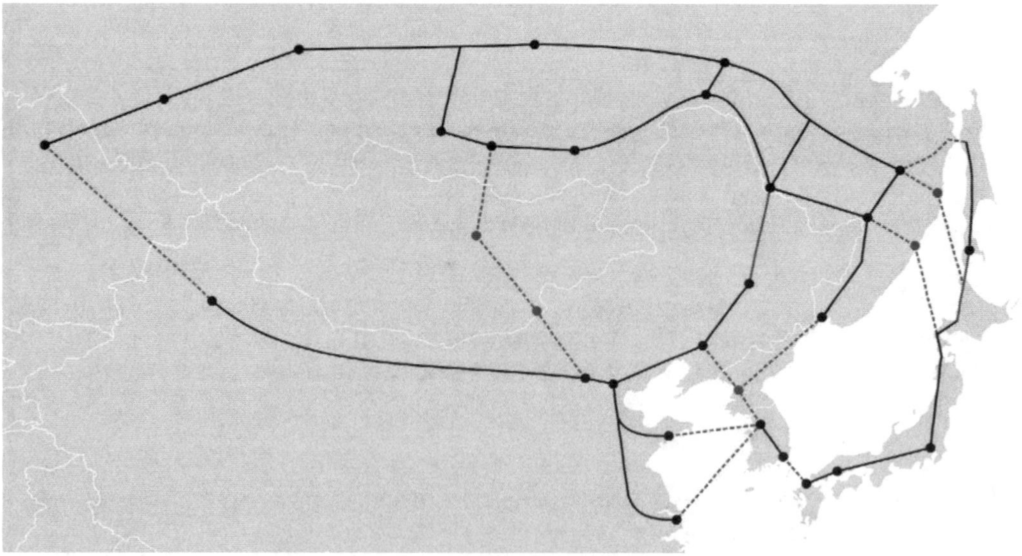

Fig. 12. Scheme of the Asian Super Grid project

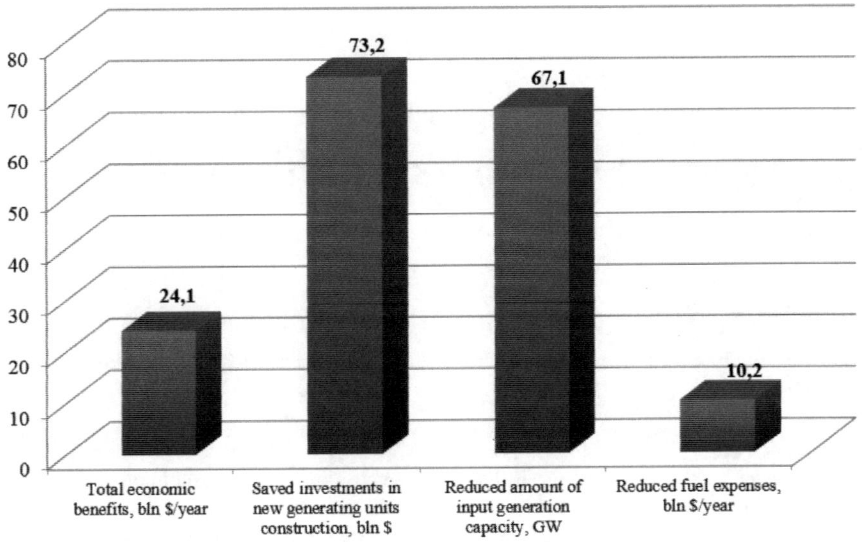

Fig. 13. Economic effects of the energy integration by 2035

Conclusion

Analysis of Northeast Asia counties' power system has been performed. Energy Connectivity data portal concept was presented. It is suggested that aggregation of power interconnection initiatives and results of research would promote power interconnection projects in Asia. Development of Northeast Asia power systems model would facilitate further research of energy connectivity issues.

References

1. Takashi Otsuki, Aishah Binti Mohd Isa, Ralph D. Samuelson, "Electric power grid interconnections in Northeast Asia: a quantitative analysis of opportunities and challenges", Energy Policy 89 (2016), pp. 311–329
2. "Gobitec and Asian super grid for renewable energies in Northeast Asia", ISBN 978-905948-143-5
3. Neil Hutcheon, Janusz W. Bialek, "Updated and validated power flow model of the main continental European transmission network", PowerTech (POWERTECH), 2013 IEEE Grenoble
4. "Yearbook of Energy Statistics", Korea Energy Economics Institute (KERI), ISSN 1226-606X, 2016
5. "Energy prices and taxes", International Energy Agency, ISSN 1609-6835, 2017
6. "Energy Balances of OECD Countries 2015 Edition", International Energy Agency
7. "Long-term Energy Supply and Demand Outlook 2015", Ministry of Economy, Japan, 2015
8. http://www.asiapacificenergy.org/ , Energy Statistisc Database, United Nation Statistics Division
9. "Analysis of power balances" System Operator report (www.so-ups.ru), Q1 2017

Interim Report by Asia International Grid Connection Study Group

Ryo Omatsu[1,*]

[1] Renewable Energy Institute, Senior Researcher, 1-13-1 Nishi-Shimbashi,Minato-ku,Tokyo,Japan

Abstract. The Asia International Grid Connection Study Group Interim Report examines the feasibility of developing an international grid connection in Japan. The Group has investigated different cases of grid connections in Europe and conducted research on electricity markets in Northeast Asia, and identifies the barriers and challenges for developing an international grid network including Japan. This presentation introduces basic contents of the interim report by the Study Group.

1 Introduction

The Asia International Grid Connection Study Group has been established in 2016 for the purpose of conducting research on international electric power networks and recommending what to do in order to open up its possibilities in Asia.

The Research Group is composed mainly of researchers of electric power systems and energy policy, experts in renewables, and people from related businesses. Since its founding in July 2016, with Renewable Energy Institute (REI) designated as secretariat, the Research Group has been discussing potentials of an international power network in Asia in a straightforward manner from multiple viewpoints. Study Group also visited Europe in January and February 2017 to inspect transmission system operators (TSOs) and government agencies there. The interim report of the research was published in May of 2017.

The main purpose of this presentation is to introduce basic contents of the interim report by the study group and help readers to understand current discussions on international grid interconnections in Japan.

2 Basic Concept of an International Power Grid

In Japan, little is known about international power grids. The country has no international connections in the first place, and few believe any such system will be built in the near future. However, looking at the world, we find cross-border power trade as common practice. Figure 1 shows how much electricity is traded between countries. The import and export ratios are defined as the proportion of imported and exported electricity to generation, respectively, both on an annual basis. In Europe as a whole, import and export ratios are 11.3% and 11.2%, respectively, a higher level of interconnection on the globe. In the region, Denmark stands

* Corresponding author: r.omatsu@renewable-ei.org

above 30% in both ratios while Belgium and Italy have high import ratio, and France and Norway have high export ratio. Despite variations between regions, electricity is tradable, and can be reasonably imported and exported with economic benefits if there is a physical interconnection, a fact that must be recognized.

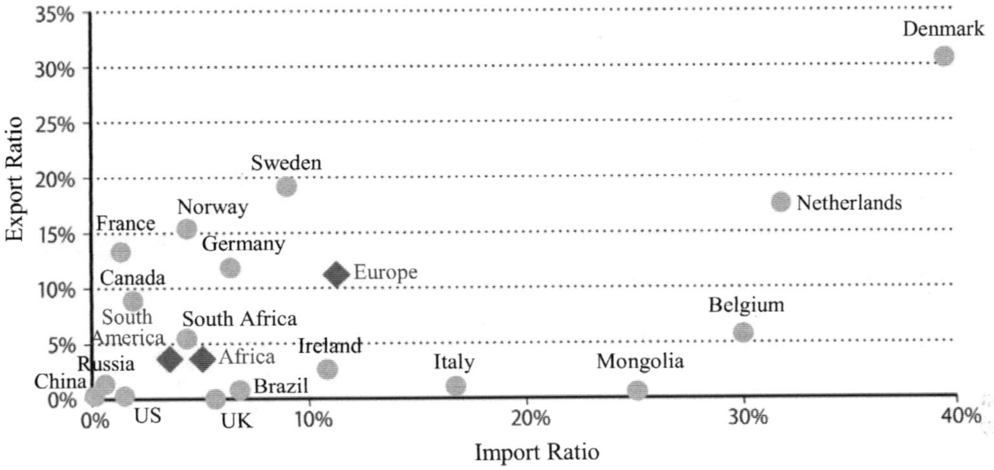

Fig. 1. Electricity import and export ratios of major countries and regions (FY 2014)
Source: IEA, Electricity Information 2016.

Why, then is the international power grid needed?

First, it leads to improvements in economic efficiency through international competition with more players, lowering the cost of electricity supply. That is, a country where electricity is supplied at a higher price could import cheaper power from another and supply it at a lower price. The greater the difference in electricity prices, the greater the benefit. This is a basic function any trade would perform. Just as is the case with ordinary goods, an electricity market, when expanded in size, becomes more economically efficient. Countries with large markets could obtain greater effect from interconnections.

That may sound as if a country originally enjoying a lower electricity rate ends up with higher price increases and suffers a loss. That is not necessarily the case. First, competitive power producers can expand into overseas markets. Export of cheap power emerges as a new business. In addition, electricity prices are subject to fine variations according to time of day and other conditions aside from average prices.

Second, inter-regional or international transmission operation contributes to a stable supply of electricity. A larger network generally makes it easier to balance supply and demand by integrating large numbers of power plants and consumers. Even with an abrupt fluctuation taking place somewhere in the network, it can respond more flexibly. A network connecting countries with different power supply structures or supply-demand patterns delivers especially great complementarity. They can also share reserve capacities.

Third, the international power grid is an effective solution to output fluctuation, and helps integrate more renewables. A larger share of wind and solar power comes with greater fluctuations of supply along with changes in weather, a challenge to grid operation Interconnection is the ultimate form of inter-regional transmission operation, and generally helps secure stability of power supply, especially when much of the supply comes from variable power sources.

3 The Current State of the International Power Grid in Europe

In the Interim Report Study Group closely studied experiences of power grid interconnections in European countries.

The history of the international power grid in Europe goes back 100 years. Among the Nordic countries, interconnection was constructed between Denmark and Sweden in 1915. In continental Europe, France, Switzerland, and Italy were connected together through 700-km interconnection in 1920. Then, after World War II, European countries started constructing interconnection on a large scale for efficient use of their limited energy resources, and international trade of electricity expanded.

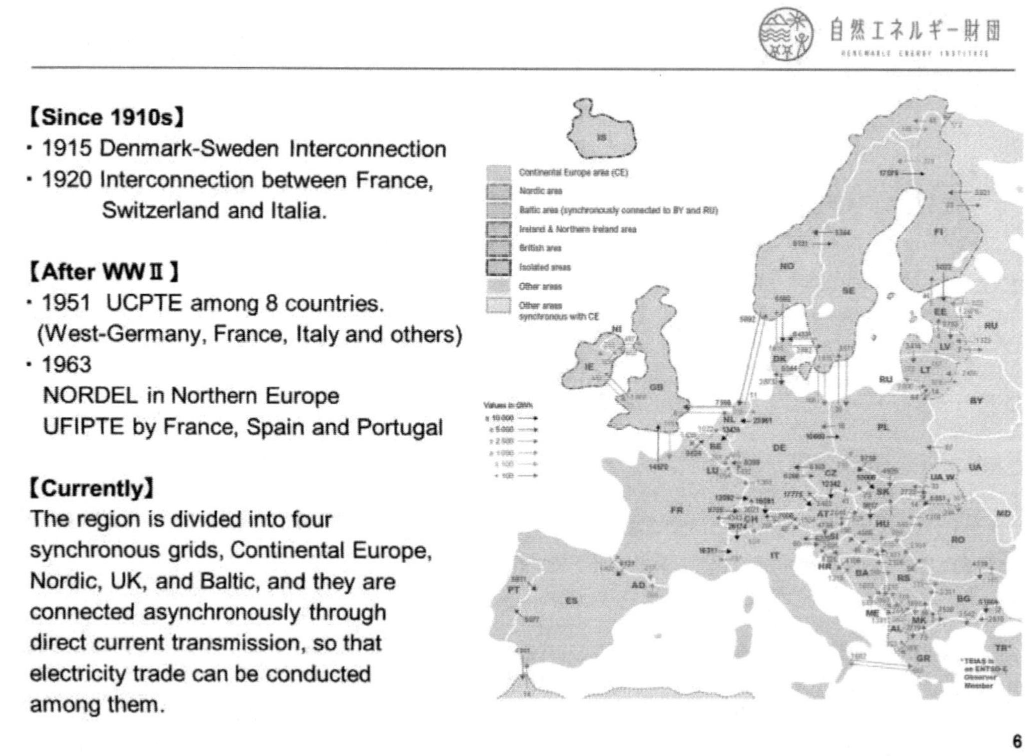

Fig. 2. Power flows between countries in Europe (2015)
Source: ENTSO-e, Statistical Factsheet 2015

Countries can expect to benefit from international power grids in different ways according to conditions they are in. Table 1 summarizes benefits the United Kingdom, Norway, Denmark, and the Netherlands each expect. Interconnection is expected to play some role when they work on challenges they face and strategic goals they set, such as high electricity price, changing power supply structure, export of electricity at lower prices, and solutions to fluctuations in renewables. Another factor that should be noted is that interconnection benefits both countries it links. For instance, Denmark and Germany, both holding a large amount of wind power capacities, can trade between them to balance supply and demand. Electricity they produce at a lower marginal cost drives down day-ahead spot prices, spreading benefits over countries around them, unless transmission lines are too congested.

 https://doi.org/10.1051/e3sconf/20182701004

Table. 1. Major benefits expected from interconnection for countries (Case of HVDC cables)
Source: Created by Renewable Energy Institute based on available data.

Country	Challenges	Major benefits expected by connected countries			
		UK	Norway	Denmark	Netherlands
UK	Higher shares of thermal and nuclear power. Replacing them in the future is a challenge. Higher wholesale electricity prices than in Nordic countries and the Continent.	–	Cheaper hydropower. NSL	Cheaper excess wind power. Viking Link	Cheaper power. BritNed
Norway	98% of electricity comes from hydropower. Hydropower is cheaper while securing capacities in drought is a challenge.	Supply capacities in drought. NSL	–	Supply capacities in drought. Cheaper excess wind power. Skagerrak	Supply capacities in drought. Supply of gas-fueled power at times when cheaper. NorNed
Denmark	Its wind power plants have a maximum capacity larger than demand at specific times. Optimizing wind power use is a challenge.	Higher adjustability. Viking Link	Cheaper hydropower. Higher adjustability. Skagerrak	–	Higher adjustability. COBRAcable
Netherlands	Electricity is supplied mainly by gas and coal-fired power plants. Higher wholesale electricity prices than in neighbors (Germany and France).	Enhanced supply capacities. BritNed	Cheaper hydropower. NorNed	Cheaper excess wind power. COBRAcable	–

4 The Current State and Feasibility of an International Power Grid in Northeast Asia

4.1 Basic conditions for future grid interconnections in NEA

From the viewpoint of development of interconnection, the greatest distinguishing feature of Northeast Asia is the fact that multiple large-scale demand zones, centers of economic activity, adjoin each other. Seen in terms of the size of economy (Table 1), Northeast Asia stretches across China, Japan, and South Korea, fairly large economies (the 2nd, 3rd, and 11th largest, respectively, in dollar-denominated nominal GDP in 2015) lying side by side. The four countries of Japan, China, South Korea, and Mongolia together account for 76% of Asia's power generation and 77% of its power consumption.

Current conditions that interconnection and electricity trading see as a given differ between countries depending on systems and institutions they have adopted, such as whether transmission companies, which operate interconnections, are separated from generation business (state of unbundling), and whether any wholesale electricity trading market is in place and works to set electricity prices.

First, each of these countries permits private-sector participation in their markets, including foreign investments, in its power-generating sector. Second, in each of these countries other than Japan, unbundling is introduced , with the grid operated by a state-run transmission company. Japan is in the process of a legal unbundling planned for completion by 2020. Third, while no international market for transactions in wholesale electricity yet exists in Northeast Asia similar to Nord Pool, Japan, South Korea, and Russia each have their own wholesale electricity trading markets as part of initiatives they are pursuing for market liberalization.

While still lagging behind Europe in electricity market liberalization, as shown above, all the countries in the region but Japan have already set up independent transmission companies, and these transmitting companies also are demonstrating proactive approaches to interconnection. In part because those companies are owned and run by state, some arrangements could be made between different regulations and systems once any agreement is formed among national governments.

Table. 2. The Current State and Feasibility of an International Power Grid in Northeast Asia.
Source: Created by Renewable Energy Institute based on data released by national governments and international organizations.

	GDP (in billion dollars) Figures in parentheses are GDP per capita (in thousand dollars)	Population (in million people)	Electricity generated (in TWh)	CO_2 emissions (in million tons CO_2)
China	8,909 (6.5)	1,376	5,811	9,154
Japan	5,986 (47.2)	127	1,036	1,208
South Korea	1,267 (25.0)	50	522	649
Mongolia	12 (3.9)	3	5	18
Russia	1,616 (11.0)	143	1,063	1,483
Northeast Asia	17,790 (~10.5)	1,699	8,437	12,512
World	74,889 (10.2)	7,349	24,098	33,508
Share of Northeast Asia	20-25%	20-25%	30-35%	~37%
Source	World Bank *Constant 2010	United Nations	BP; For Mongolia, figure from IEA in 2014	BP; For Mongolia, figure from IEA in 2014

There are large differences between countries in price level and power demand patterns, which creates good conditions for mulual comprementarities achievable through grid interconnection.

4.2 Rapid expansion of Renewable Energy as a driver for interconnections.

In North East Asia each country has been actively promoting investments in domestic renewable energy. Against the backdrop of serious air pollution issues and remarkable decrease in the costs of renewable energy installation and operation, this movement is expected to be further accelerated.

According to future forecasts announced by international agencies for 2030-2050, in 2030 it is projected that 30-50% of electricity generated in each country of Northeast Asia will be supplied from renewable energy (Figure 3). In terms of the period up to 2050, there is a report (NDRC) forecasting that renewable energy will account for 80% of power generation in China.

In considering a potential for power supply by renewable energy in Northeast Asia, Mongolia is particularly an important country. The potential for wind power and solar PV mainly in the area around the Gobi Desert in the south has been highly valued.

IRENA's "Mongolia Renewables Readiness ASSESSMENT" (March 2016) estimates RE potential in Mongolia as 10,673 TWh for wind power and 4,777 TWh for solar PV. The total potential power generation will be 15,000 TWh or more, which is much more than the total electricity demand in China (5,693 TWh) and Japan (949 TWh) in 2015. It is expected that Mongolia will supply power to neighboring countries in the future by utilizing its rich renewable energy.

As above mentioned, Northeast Asia is a region with not only major economies adjacent to each other but also a high potential for power generation by renewable energy. In order to make the most of renewable energy that has been developing in Asian countries, grid flexibility for power interchange across Asia is required, and the need of an international power grid is increasing.

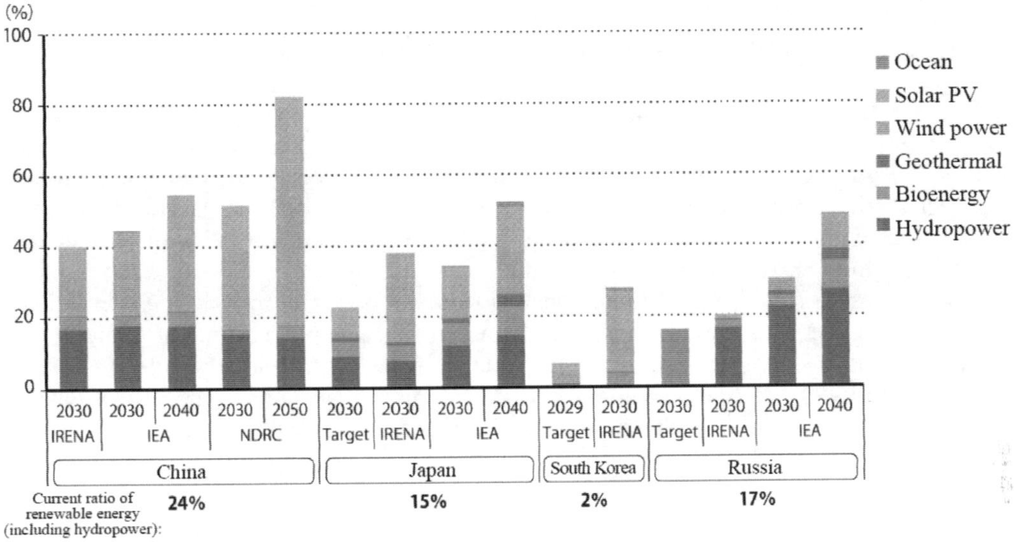

Fig.3. Future prospects of power supply structures in Northeast Asia (2030-2050)
Source: Created by Renewable Energy Institute based on IRENA (REmap 2016), IEA (WEO 2016 - 450 scenario), NDRC (China RE Outlook 2016 - High Penetration scenario)

4.3 International power grid schemes surrounding Japan: From concept to the real plan

For last decades, different research institutes and organizations have proposed it's own concepts for Power grid interconnection in North East Asia.

Since 2011, international power grid concept has been proposed mainly by power and transmission companies in Northeast Asia. Inspired in part by the Asia Super Grid vision (promoted by Renewable Energy Institute), major transmission companies in Asia start to formulate their own international power network scheme in Asia in conjunction with the vision.

For example, the Korea Electric Power Corporation announced the Northeast Asia Super Grid plan in 2014 in cooperation with the Russian research institution Skoltech and other organizations. The company set out the concept of a Smart Energy Belt that connects Japan, China, South Korea, Russia, and Mongolia with a highly efficient electricity supply-demand system combining power storage technologies and smart grid

In 2015, the State Grid Corporation of China (SGCC), the world's largest power transmission company, announced its Global Energy Interconnection (GEI) vision. In March 2016, the international nonprofit foundation Global Energy Interconnection Development and Cooperation Organization (GEIDCO) was established with the goal of realizing this GEI concept. Consisting of research institutions and relevant enterprises in various countries, GEIDCO aims to develop a global power grid to utilize renewable energy.

It also is worth noting for Japan that the Russian government has formally made a specific proposal on interconnections between the two countries. In September 2016, the Eastern Economic Forum was held in Vladivostok in the Far East. At its plenary session, President Putin said, "we support the initiative of Russian, Japanese, South Korean, and Chinese companies to create an Energy Super Ring linking our countries." He also proposed setting up an intergovernmental working group for the initiative.

In this way, enterprise-driven movements towards the realization of the Asia international power grid scheme have been accelerated since 2016.

5 Legal considerations to work on international transmission in Japan

Since the legal structure on electricity business in Japan did not envision interconnection, there is a high likelihood for necessity of consideration on legal and regulatory reforms. Although a wide variety of matters need to be examined, we will clarify the issues focusing on the license for interconnection business, and regulations on installation of undersea transmission lines and electricity market trading.

The first topic is the license for interconnection business. When considering international power grid connection as having the same characteristics as regional grid connections, it is conceivable that the Transmission Business license under current law (Article 27-4 of the Electricity Business Act, EBA) could be used. As the current law does not specify provisions on connection to overseas, these provisions can be added. The Telecommunications Business Act (TBA) can be used as a reference. This law calls for licensing of telecommunication enterprises regardless of whether they are international or domestic (Article 9 of TBAe). Its provision specifies that a telecommunications enterprise shall obtain authorization from the Minister of Internal Affairs and Communications when it intends to conclude an agreement on important matters related to telecommunications operation with foreign governments or foreign companies (Article 40 of TBA, Article 27 of the Ordinance for Enforcement of TBA).

Another method is to establish new licenses especially for interconnections. For example, we can add a new section titled "International transmission business" to specify the details of the license.

Among other countries, there are examples of both those that specify explicitly the differences between international transmission licenses and standard transmission licenses under the law (England and Wales, Ireland, etc.), and those that do not (Northern Ireland, France, United States, Canada, etc.). In the latter case, some specify the operation of international transmission business as a special case in the details of the individual license, and some require authorization from the government for installation of interconnections and international electricity trading. Various forms of regulations can be considered, depending on the existing systems or policies of the country.

Second, as for regulations on installation of undersea transmission lines, EBA has no specific provisions concerning installation of undersea transmission lines. Each country may install undersea cables freely in extraterritorial waters (refer to Articles 57, 79, and 87 of the United Nations Convention on the Law of the Sea). Since the waters between Japan and Russia and South Korea, anticipated as partners for installation of such cables, consist only of each country's exclusive economic zone, once agreements are concluded with these countries, there would be no particular problems with their installation under the treaty. Also, many fiber-optic cables for telecommunication are installed in the seas around Japan (Figure 4).

And in fact, multiple undersea power cables have been installed in the seas of Japan. Therefore, nonexistence of the provisions would not be impediments to the installation. However, the Fishery Act does have provisions concerning installation of underwater cables (Article 39(1)). As a reference, TBA has provisions on coordination with fishing rights and protection of submarine cables (Articles 140 through 143). Also, the Wire Telecommunications Act prohibits the installation of unauthorized international

https://doi.org/10.1051/e3sconf/20182701004

communications cables (Article 4). Therefore, it is worth considering that EBA would include similar provisions.

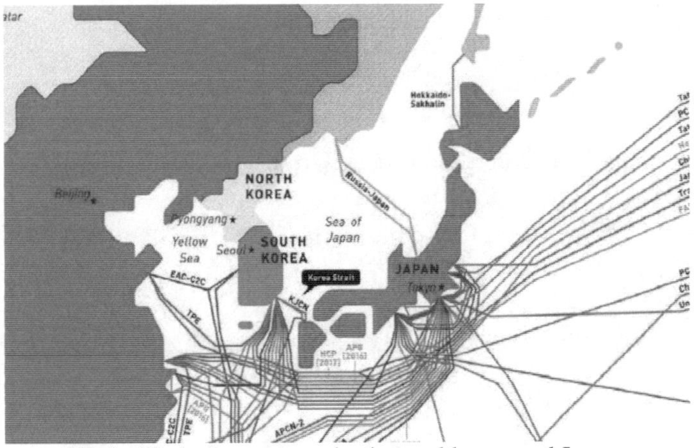

Fig.4. Undersea communications cables around Japan
source : TeleGeography Submarine Cable Map 2016.

Third, it is necessary to ensure participation in electricity markets from overseas. Now we would like to examine whether the Japan Electric Power Exchange (JEPX) needs some amendments to current market rules to allow international trading of electricity. If Japan imports electricity from overseas, foreign power generation companies or power trading companies may participate in the market as sellers. If Japan exports electricity to overseas, foreign retailers or power trading companies may participate in the market as buyers.

Conclusion

As we have discussed so far, an international power grid is economically reasonable in general. Also, compared with domestic grids, there is no particular technical difficulty in an international power grid. These aspects have been proved by the long history of the international power grid in Europe. Moreover, interconnections are utilized not only in Europe but also in Southeast and South Asia. In light of global renewable energy expansion in the future, many interconnection construction projects are planned across the world. We can say that there are no absolute reasons that it would be impossible to construct or operate international power grids in Northeast Asia including Japan.

Actually, there may have been little need for an international power grid in Japan or Northeast Asia in the past. However, now that China has the largest installed capacity of renewable energy in the world and Russia and Mongolia have huge potential, we should change the recognition of the continuing situation where Japan remains unconnected with any neighboring countries. We believe that the 21st century should be the age of electricity trading not only in Northeast Asia or all over Asia, but also on a global basis. The fact that other countries of Northeast Asia, as well as their transmission and power companies, have made proposals for international power grids connecting this region should be given serious consideration.

The time has come for us to break through the preconceived notion that as an island nation an international power grid would be impractical in Japan and to recognize that such a network is essential for both the Japan of today and other countries in Asia and could, ultimately, contribute to resolving global energy issues.

At the same time, the most important challenge for Japan to launch a specific international power grid project is the determination as a nation. In order to bring the project into practice, the government must have a firm commitment to the realization of international transmission of electricity and begin formal discussions with other countries' governments.

Reference

1. Asia International Grid Connection Study Group, *Interim Report* (2017)

The role of hydropower in electric power integration of Asian countries

Lev Belyaev, *Vladimir* Savelyev, and *Lyudmila* Chudinova*

Melentiev Energy Systems Institute, Siberian Branch of the Russian Academy of Sciences, Department of Electric Power Systems. 130, Lermontov str., Irkutsk, Russia

Abstract. The possible role of hydropower plants in formation and operation of interstate power pools is described based on the generalization of the world experience. Peculiarities of the influence of hydropower on development of interstate electric ties in this part of the world and potential effects are showed on the example of Central and Northeast Asia.

1 Introduction

Hydropower plants (HPPs) as sources of renewable energy produce cheap and environmentally friendly electricity and have high regulating and maneuvering capabilities. Pumped storage power plants (PSPPs) are the most inexpensive high-power energy storage. This fact determines a special role of hydropower engineering in development of interstate electric ties (ISETs) and formation of interstate power pools (ISPPs) in many regions of the world. In particular, the construction of HPPs contributed to the creation of ISPPs in the territories of North, Central and South America, Scandinavia, Central and South Africa [1]. Hydroelectric power plays an important role in the projects of integration of the electric power sectors of the countries of Central, Northeast, Southeast and South Asia. Table 1 characterizes the hydropower potential and the state of its usage in these regions [2-4, etc.).

Construction of HPPs and PSPPs within the interstate power pool replaces fossil-fueled power plants from their energy balances, and promotes the development of renewable energy sources (RESs), mitigating the environmental problems in the participating countries. The regulated hydropower plants maintain uniform operation of nuclear power plants and can compensate for natural fluctuations in electricity production of solar, wind and other renewable energy sources, improving thus reliability and efficiency of power supply in power pools. In some cases, construction of ISPPs contributes to solving the problems of using water and hydropower resources of transboundary water bodies (rivers and lakes).

On the other hand, the increasing scales of ISPP as a unified power system makes it possible to increase the installed capacities of hydropower plants and improve their cost effectiveness.

The current problems and possible directions for the development of interstate power pools in Central and Northeast Asia are considered below in the context of the impact of hydropower of countries with a high potential of hydropower resources on them.

* Corresponding author: chudinova@isem.irk.ru

Table 1. State of hydropower resources.

Region	Installed capacity (including pumped storage) in 2016, GW	Generation in 2016, TWh	Gross theoretical capability, TWh/year	Technically exploitable capability, TWh/year	Current utilization
World total	1239.5 (149.7)	4102	43378	15517	26%
South America	161.1 (1.0)	709	7893	2807	25%
North and Central America	200.9 (22.6)	702	7601	1842	38%
Europe	223.0 (50.5)	595	3129	1199	50%
Africa	33.6 (3.4)	106	4380	1484	7%
Asia and Pacific including:	620.9 (72.2)	1990	20375	8185	24%
Afghanistan	0.442	1.17	394.5	88	1%
China	332.11 (26.7)	1180.7	6083	2474	48%
Iran	11.278 (1.0)	18.2	448	179	10%
Japan	50.035 (27.6)	92	718	135.8	68%
Kazakhstan	2.372	9.27	198.6	61.9	15%
Korea, North	5.000	12.040	50	> 26.0	46%
Korea, South	6.738	6.622	51.8	26.4	25%
Kyrgyzstan	3.091	13.32	163	99	13%
Mongolia	0.026	0.059	56.2	22	0%
Pakistan	7.115	34.272	>475	219	16%
Russia, including:	51.45 (1.4)	186.6	2395	1670	11%
Sibir	26.5	103.2	993	757	14%
Far East	5.8	18.6	1009	684	3%
Tajikistan	5.19	16.9	527	317	5%
Turkmenistan	0.001	0.003	23.9	4.8	0%
Uzbekistan	1.889	10.59	88.5	27.4	39%

2 Central Asia

Central Asia includes 6 countries: Iran, Kazakhstan, Kyrgyzstan, Tajikistan, Turkmenistan and Uzbekistan. In addition, as will be shown below, Afghanistan and Pakistan, geographically belonging to South Asia, are closely related with them in the field of electricity supply and prospects for the establishment of a regional ISPP.

The countries that previously were members of the interconnected power system of Central Asia of the Unified Power System of the USSR play an important role in the electric power integration of Central Asia countries. After its disintegration, these countries created their own ISPP - Central Asian Power System (CAPS) [1]. This pool allowed preservation of all the advantages of energy integration that were achieved owing to the centralized management of development and functioning of the electric power industry of the Soviet Union. In particular, this made it possible to use rationally diverse energy resources of these countries: coal in Kazakhstan, oil and natural gas in Turkmenistan, hydropower in Tajikistan and Kyrgyzstan. Besides, the unified integrated water and energy complex was created and managed from the joint center in Uzbekistan. It ensured the balance of seasonal fluctuations in electricity demand and water requirements for irrigation purposes with fluctuations in its inflow to the HPP reservoirs.

However, the political and trade contradictions between individual Central Asia countries ended in the exit of Turkmenistan first (2003) and then Uzbekistan (2009) from the electric ring. This situation gave rise to the problems of meeting the irrigation requirements by the Tajik and Kyrgyz hydropower plants in Uzbekistan, on the one hand, and, on the other, maintaining the electricity balances in Tajikistan and Kyrgyzstan. The problem is that the loss of hydro energy in the summer season that is caused by additional water discharge because of the low electricity demand leads to its shortage in the winter season [5, 6]. These hydropower losses amounted to about 6 TWh per year. In addition, the construction of the Rogun HPP on the Vakhsh river and the Sangtuda HPP on the Naryn was halted.

To date, the situation in the electric power industry and water management of the region has changed significantly, in particular, owing to the active participation of the UN and the world financial system (World, Asian, Islamic and other banks and funds) [7, 8]. The International Conference under the auspices of the United Nations "Water for Life", held in 2015 in Dushanbe [5], which highlighted the importance of these problems for the socio-economic development of the countries of the region, also served to this change.

At the end of 2017, Tajikistan and Uzbekistan proposed restoration of the unified regional power system, taking into account, in particular, the need to solve the problems of electricity supply to neighboring Afghanistan [8]. This makes it possible to raise the issue of creating interstate electric lies between CAPS, Afghanistan and Pakistan once again.

The project of expanding the CAPS borders due to the availability of unused seasonal hydropower in the late 1990s - early 2000s, as well as the potential for a significant increase in its production at the new HPPs was developed on the initiative of the World Bank for Development ("CASA-1000" project) [5,6] . Initially, the project envisaged completion of the construction of the Rogun HPP and the Sangtuda HPP and in addition, construction of new hydropower plants (the Vakhsh cascade in Tajikistan, the Kambarata HPPs and the Toktogul cascade in Kyrgyzstan). In this case, the export potential of these countries could increase to almost 40 TWh by 2025. For the delivery of this electricity to Afghanistan and Pakistan, it was planned to build transboundary transmission lines (TBTLs) with a voltage of 500 kV: "Kyrgyzstan-Tajikistan-Afghanistan" with a converter substation in Kabul and the direct current TBTL to Peshawar (Pakistan). The project also included proposal about construction of a 500 kV TBTL from Turkmenistan to Iran.

The implementation of CASA-1000 is the first stage in creating the Central Asia-South Asia Regional Electricity Market (CASAREM) (Fig. 1) [9]. The project can promote the development of electricity trade between the countries of the two regions of South and Central Asia and involves the export of excessive summer electricity from Tajikistan and Kyrgyzstan to Pakistan. In this case, Afghanistan can make a profit as an electricity transporter.

In total, there are plans to create an asynchronized power system of 6 South and Central Asia countries (Turkmenistan, Uzbekistan, Tajikistan, Kyrgyzstan, Afghanistan and Pakistan) beyond 2020, based on the existing and planned AC electric networks in Afghanistan.

In addition, through the power systems of India Pakistan is developing ISETs with the power systems of the South Asia countries belonging to the South Asian Association for Regional Cooperation (SAARC).

Thus, the main incentive for interstate energy integration of the countries of Central and South Asia is the use of the currently unclaimed electricity generation at HPPs, the possibilities of its additional increase at new HPPs (under construction and planned), and the expansion of natural gas export for its use at export-oriented thermal power plants (TPPs).

3 Northeast Asia

This Asian region includes 6 countries (China, the Republic of Korea, the DPRK, Japan, Mongolia and Russia). At the same time, the electric power integration within the framework of the future interstate power pool of Northeast Asia mainly concerns only the North and Northeast China and the eastern regions of Russia

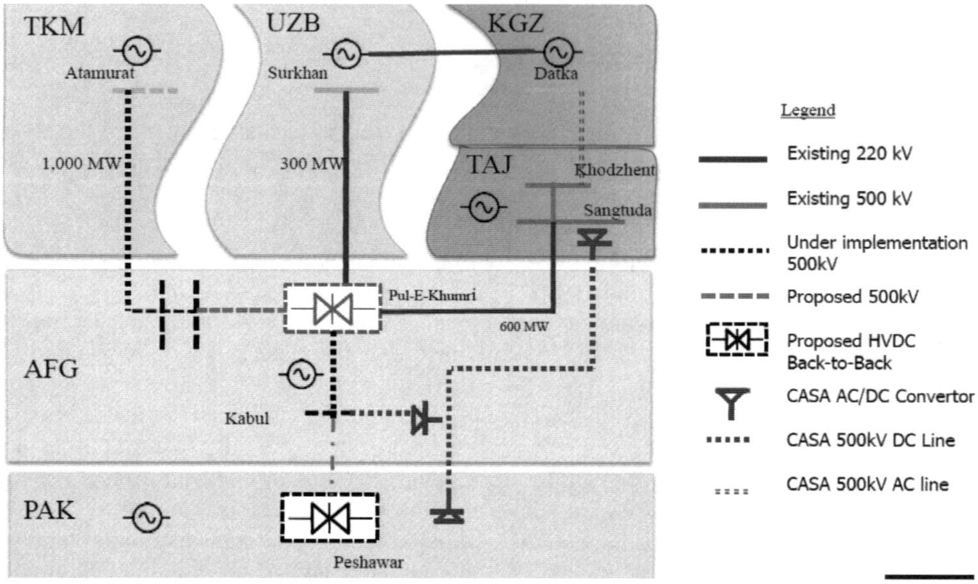

Fig. 1. Central Asia – South Asia Regional Electricity Market (CASA – 1000 and TUTAP).

(Eastern Siberia and the Russian Far East). Interest in the development of ISETs and the establishment of ISPP NEA has a long history. However, despite favorable conditions [10] and high potential efficiency [11], its formation is still at the stage of the pre-feasibility study. In all related research, special attention is paid to hydropower.

This is due to a number of reasons. First of all, the two countries (China and Russia) have the greatest potential in the world for hydropower resources (Table 1). Japan and the DPRK also have great hydro potential. Only the Republic of Korea and Mongolia are poor in hydropower.

Further, some countries of the region have faced acute environmental problems and are forced to actively develop renewable energy.

Then, the countries under consideration are characterized by different conditions for hydropower development in the long view: the rates of economic development and the growth of electricity consumption, as well as the need for new power plants, the structure of generating capacities and the fuel balance of TPPs.

And finally, different countries have different investment opportunities and pursue their own policy in construction of HPPs with a long payback period.

It should also be noted that there is no generally accepted project for the ISET development in the region and the establishment of the regional ISPP of Northeast Asia. Currently, the main focus is on the choice of priority interstate electric ties.

All these factors influence the HPP participation in the electric power integration of individual countries in the region and the prospects for their hydropower development, taking into account external electric ties. This is especially true of the eastern regions of Russia (Fig. 2).

Let us dwell on the specific features of hydropower development in the NEA countries and their possible role in the formation of the regional ISPP.

As can be seen from Table 1, Japan has practically utilized its hydropower resources, including the construction of pumped storage power plants. Their rapid development until 2011 was necessitated by the need to level the electricity production of intensively constructed nuclear power plants. After the accident at the Fukushima-1 nuclear power plant and decommissioning of practically all nuclear power plants in the country, the available capacities of PSPPs provided opportunities for the large-scale construction of

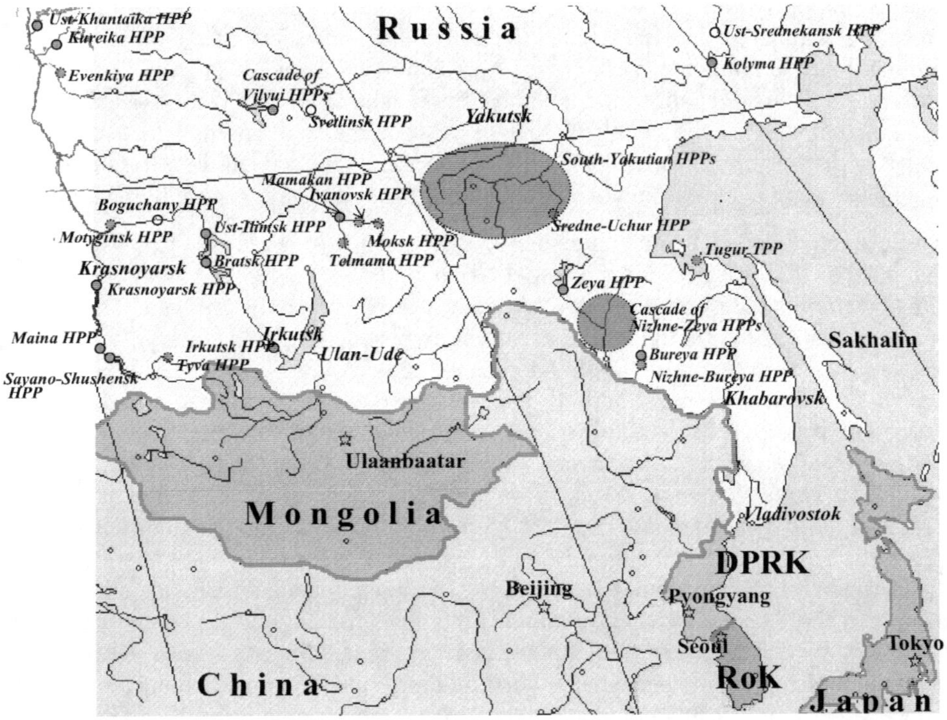

Fig 2. Map of hydropower **resources** development in Eastern Russia and directions of ISETs with NEA countries.

solar power plants. The need of this country to develop interstate electric ties with the eastern regions of Russia rich in hydropower resources is explained at the present stage by the necessity of increasing the import of environmentally clean electricity. In this case its generation by thermal power plants in Japan will decrease, which will allow the environmental problems to be solved and the dependence on the import of hydrocarbon fuels to be overcome. Therefore, Japan is interested in importing clean electricity from Russia.

The **Republic of Korea** experiences similar problems and needs in the development of interstate electric ties in the context of hydropower development.

In the **DPRK**, the hydropower plants, in the absence of the possibility of using other energy resources, in particular coal, are practically the only source of electricity. Part of their generation was even exported to the Liaoning province of Northeast China. Therefore, this country is interested in diversification of electric power sources, which, if the political situation changes, include ISETs with China and Russia.

Mongolia's electric power industry currently relies on coal-fired cogeneration plants, leveling the electric load curves, in particular, due to the reverse TBTL from the Gusinoozerskaya TPP to Buryatia. In order to cover the need for maneuvering capacities, the country plans to construct several hydropower plants on the Selenga river and its tributaries. These plans aroused sharp criticism of the Russian environmental community because of the possible adverse impact on the ecosystem of lake Baikal. However, this issue is ambiguous. Based on the current world practice, the use of water resources of the transboundary rivers is, in principle, allowed. However, in this case we are talking about the use of water reserves of the object of global significance according to UNESCO classification. For a time, the problem of power shortage in Mongolia can be solved by increasing exports from the Gusinoozerskaya TPP. But this is not a cardinal decision. Therefore, additional research is needed to resolve the contradiction that has arisen.

Another line of interstate electric ties between Russia and Mongolia consists in the construction of the TBTL from the Sayano-Shushenskaya HPP to the Altai-Uliastai power system in the west of the country. This ISET will make it possible to use the locked capacity of the Sayano-Shushenskaya HPP, ensure the development of the mining industry in this power system and the creation of the unified national power system of Mongolia.

People's Republic of China. The role of this country in the electric power integration both in the Northeast region and the whole Asian continent is largely decisive because of its leading position in their economic development.

China is ranked first in the world in electricity production and consumption, has a great investment, construction and technological potential. At the same time it is engaged in intensive hydroelectric construction and is ready to render any assistance in the construction of new HPPs in other regions of the world, including Russia.

The level of practical use of hydropower potential in China in 2016 reached 48%. The constructed "Three Gorges" HPP with an installed capacity of 22.4 GW is the world's largest HPP. The construction of another giant hydropower plant Baihetan with a capacity of 16 GW, an average annual output of 60.24 billion kWh has been started.

In addition to solving domestic problems, the development of China's hydropower, in particular in the Yunnan province, promotes the development and functioning of interstate electric ties in Southeast Asia. The Yunnan province with its largest rivers Yangtze, Mekong and Nu is one of the richest world regions in hydropower resources [12]. In the province there are 128 hydropower objects with a total capacity of 16.5 GW. They can cover the increasing electricity consumption not only in the Southern power pool of China, but also in the whole Greater Mekong region. Thus, the hydropower industry of this province can contribute to the formation of the unified ISPP as an Asian part of the Eurasian electric power space [13].

It should be noted that China reached a very high technological and economic level in other branches of the electric power industry and the electrical engineering. The manufacture of relatively cheap equipment with ultrasupercritical steam parameters for thermal power plants, electric grid equipment with a voltage of 1100 kV AC and ± 800 kV DC are the vivid examples. The wind and solar power engineering are also actively developing in the country. This reduces the need and competitiveness of electricity imports from other countries of the region.

At the same time, the problem of developing interstate electric ties in the region and participation of China in this process remains relevant.

4 Eastern regions of Russia

Eastern Siberia and especially the Russian Far East, have the richest unused hydropower resources. They served as the basis for active hydropower construction in these areas in the 20th century. As a result, the unique Angara-Yenisei cascade of HPPs, Vilyuiskaya, Zeyskaya and Kolyma hydropower plants were constructed on the same rivers. This provided a high share of hydropower in the structure of generating capacities in IPS of Siberia and IPS of East. Meanwhile, the development of the hydropower potential of the eastern regions is still lagging behind its utilization in the European part of Russia.

As is shown in [14-16], currently the possibilities of hydropower construction in Russia and in its eastern regions have changed significantly due to a decreasing need for new sources of electricity, limited investment opportunities for the state and private companies, difficulties in economic development and development of natural resources of new territories, rise in the cost of HPP projects due to relocation of construction sites to remote, poorly developed territories, etc.

In these conditions, the development of interstate electric ties with neighboring countries, primarily East Asia, is of particular importance. Hydropower in the eastern regions of Russia can participate in expanding these ties in several directions (Fig. 2).

Firstly, there are many purely export projects for the construction of HPPs on the Siberian and Far-Eastern rivers in order to supply their electricity to China, Japan and the Korean peninsula [9,10]. In particular, the project of electricity exports to Japan from the HPPs of the South Yakutia hydropower complex [15].

Secondly, the hydropower plants in the eastern regions of Russia, which have large water reservoirs, primarily the Angara and Yenisei reservoirs, play an important role in the implementation of integration (systemic, synergetic) effects of the creation of ISPP NEA. In particular, this applies to the effect of the discrepancy in the seasons of the annual maximum loads, on the one hand, in IPS of Siberia and IPS of East, and, on the other hand, in the power systems of Japan, the Republic of Korea and North China. Realization of this effect makes it possible to considerably reduce the commissioning of generating capacities in these countries. At the same time, the fuel costs of thermal power plants will be saved and the emissions of pollutants into the environment will decrease.

Thirdly, the studies of the effectiveness of creating the ISPP NEA with the help of the mathematical model ORIRES revealed an additional integration effect associated with the use of the regulating capabilities of existing and new Siberian and Far Eastern hydropower plants at the control of its short-term operation conditions [11]. It consists in regulating the variable electricity production by renewable energy sources in neighboring countries.

On the whole, the studies of the Energy Systems Institute showed an important role of hydropower plants in the eastern regions of Russia and other countries in the formation of ISPP NEA and realization of the effects achieved in development of their hydropower engineering will be possible only with creation of the considered interstate interconnection. This conference demonstrated the interest in this ISPP.

The paper did not consider the impact of hydropower engineering on the development and operation of ISPPs in two other Asian regions: the South and Southeast regions. This issue requires special studies taking into account the existing and future electric ties of these regions with the countries of Central and Northeast Asia, as well as their role in formation of the Pan-Asian super power pool.

References

1. L.S. Beliayev, S.V. Podkoval'nikov, V.A. Savel'yev, L.Yu. Chudinova, *Efficiency of Interstate Electric Ties* (2008)
2. World Energy Resources, (2013). Available: http://www.worldenergy.org [Accessed 16 April 2016]
3. 2017 Hydropower Status Report, (2017). Available: https://www.hydropower.org [Accessed 20 August 2017]
4. Hydropower generation, (2014). Available: http://www.unesco.org [Accessed 30 August 2016]

5. V. Pervukhin, Energy Policy, **1**, 26-36 (2016)
6. R. Sharma, *The Strategy of World Bank for Support of Energy Projects in Central Asia* (2005)
7. Central Asia May Restore The Unified Energy System, (2017). Available: http://easttime.info [Accessed 20 May 2017]
8. Uzbekistan is interested in widening the Central Asian Ring, (2017). Available at: http://news.tj [Accessed 02 October 2017]
9. Central Asia - South Asia Regional Energy Trade (2014). Available: http://www.carecprogram.org [Accessed 02 October 2017]
10. L. Yu. Chudinova, S.V. Podkovalnikov, V.A. Savelyev, Studies on Russian Economic Development, **26** (4), 403-412 (2015)
11. S.V. Podkovalnikov, V.A. Savelyev, L.Yu. Chudinova, Energy, **5**, 16-32. (2015)
12. T. Hennig, W. Wang, D. Magee, D. He, Water 2016, **8**(10), 476 (2016)
13. V.A. Savelyev, L.Yu. Chudinova, Energetic, **7**, 33-36 (2014)
14. L.S. Belyaev, N.I. Voropai, S.V. Podkovalnikov et al., Energy Policy, **1**, .26-36. (2016)
15. L.S. Beliayev, S.V. Podkoval'nikov, V.A. Savel'yev, L.Yu. Chudinova, *Procceedings of the Proceeding of 6th International Conference — Asian Energy Cooperation: Forecast and Realities* (Irkutsk, Russia, 2008)
16. Podkovalnikov S.V., Savelyev V.A., Chudinova L.Yu. Northeast Asia Energy Focus, **5** (3), 36-43 (2008)

Mongolia's potential in international cooperation in the Asian energy space

Sereeter Batmunkh[1], *Valery Stennikov*[2,*], *Bayar Bat-Erdene*[3], *and Altay Erdenebaatar*[4]

[1,3,4] Mongolian State University of Science and Technology

[2] Melentiev Energy Systems Institute of Siberian Branch of the Russian Academy of Sciences, 130, Lermontov str., Irkutsk, 664033, Russia

ABSTRACT. The paper is concerned with the issues of interstate electric power interconnections to be created in the countries of Northeast Asia. The conditions are formulated, the problems are stated, and solutions for Mongolia's entry into the Asian energy space are proposed. The electricity consumption rates are growing, however, the Northeast Asia countries differ considerably in available energy resources to cope with this growth. Therefore, the need to build international electric power interconnections that take into account climatic features, seasonal peak load differences and other factors in order to rationally match power demand and supply is getting increasingly more obvious. Mongolia can take an active part in this process, as the country is rich in energy resources and interested in their development to meet their domestic needs and exchange with neighboring countries. The establishment of interstate power interconnections in the Northeast Asia countries represents a topical task whose solution will make it possible to meet the demand of this region for electricity on mutually beneficial terms. Mongolia has a good spatial position, energy resources and is interested in ensuring domestic energy balance. Therefore, the country can be an active participant in such an integration process.

1 Introduction

In light of noticeable differences in the levels of general development of the countries, availability of renewable (wind, solar, hydro) and traditional fossil energy resources, the issues of electricity import and export have become an indispensable part of energy policy of many countries in the world. Therefore, energy market is expanding, first of all, to the neighboring countries. The principle of energy market expansion is based, as is known, on the construction and development of interstate power interconnections that can be exemplified by the following currently operating interconnections: [1]

1. Eastern and Western Grid in the USA.
2. ENTSO-E (European network of system operator in electric power industry in Europe.
3. Unified power system/Interconnected power system of NIS and Baltic Countries.

[*] Corresponding author: sva@isem.irk.ru

4. Greater Mekong electricity subregional market.
5. SAARC (South-Asian Association for Regional Cooperation) Power Grid in Asia.

An interstate power interconnection to be created as a single technical system offers good possibilities of ensuring mutually beneficial power trade, reliability of uninterrupted power supply to the regions, increase in the efficiency of installed capacities at various power plants, technological advance of energy infrastructure, creation and expansion of green electricity generation and solving environmental problems, and construction of large-scale joint power facilities in the member countries of this interconnection.

An analysis of the international experience shows that interstate power interconnections tend to be constructed in a limited territory, including two and more states or in the off-shore zones.

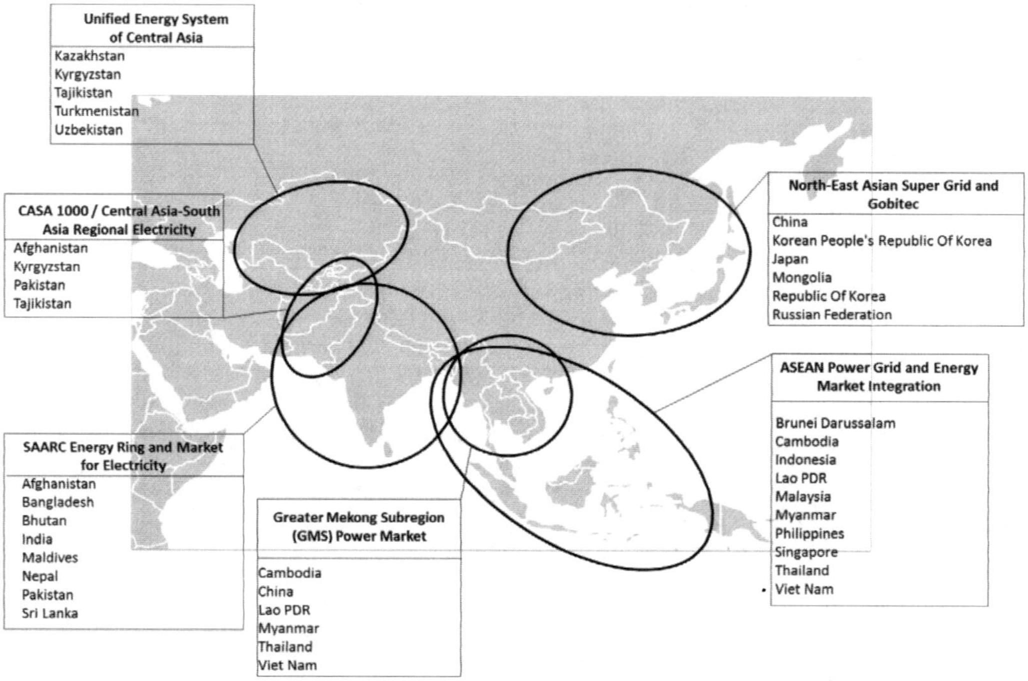

Fig. 1. Some projects on interstate power interconnections in Asia and Asia-Pacific region.

The studies on the construction and development of the interstate power interconnections are mainly aimed at the creation of large capacities involving renewable energy sources (solar, wind and hydro) to eliminate shortage of generation capacity of conventional energy sources, on the one hand, and to provide favorable environment, on the other hand. For example, China is building solar and wind integrated systems, Mongolia is doing research on the construction of such integrated systems in the Gobi desert [2]. These systems are further planned to be part of the interstate power interconnection in Northeast Asia – Asian Super Grid (Fig. 2) [3].

2 Potential of Mongolia's participation in the energy cooperation

In order to take part in the interstate power interconnection in Northeast Asia (Asian Super Grid), Mongolia can undertake the construction of high-power power plant on the basis of available renewable energy sources in the Gobi desert, and construction of high-voltage AC and DC transmission lines, which will ensure stable energy development. It is worth noting that there are other countries interested in participation in this project. These are Russia, China, Japan and South Korea. Currently, serious studies are being conducted to determine the development prospects for this region [3, 4]. Some of these countries border one another, have long experience of successful trade cooperation, have sufficient investment and financial capabilities to construct large heat, electricity, hydro and nuclear power plants. Russia jointly with China and Japan, and China with the Democratic People's Republic of Korea started the implementation of the first joint large-scale energy projects [5-8] which can underlie the creation of the interstate power interconnection in Northeast Asia. The possibility of participation and role of Mongolia in the project "Asian Super Grid" [9, 10] cause the need to carry out scientifically-grounded research aimed at determining the sites for the new generation capacities, considering spatial distribution of energy resources in the regions and situation in the electricity markets.

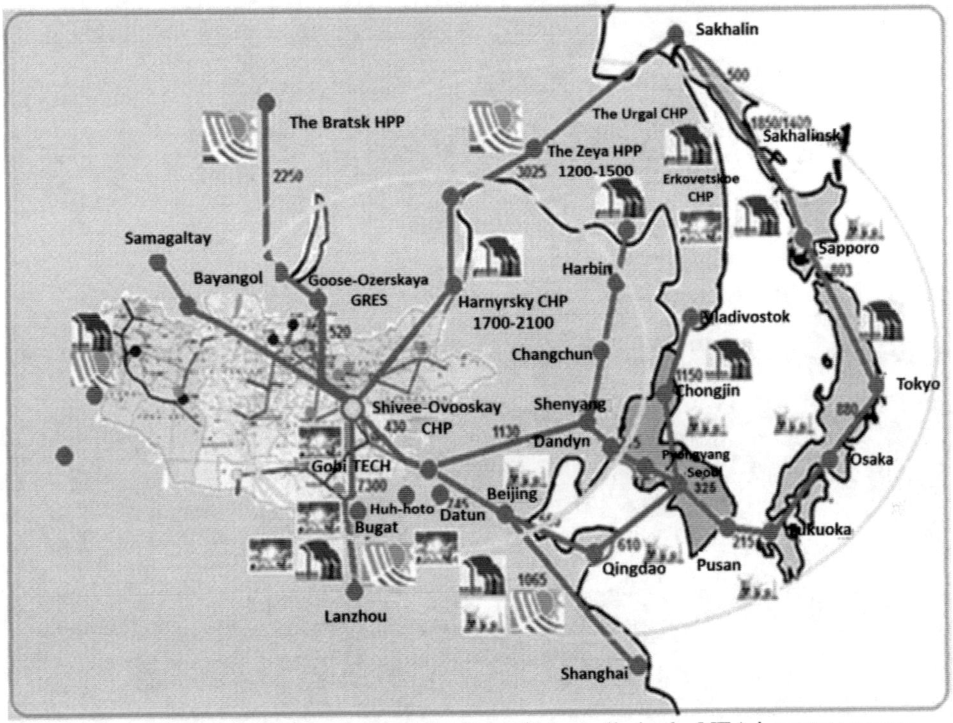

Source: A preliminary report on the participation of Mongolia in the NEA interstate power interconnection. National RES Center. 2015.

Fig. 2. A general scheme of the proposed interstate power interconnection of Northeast Asia – "Asian Super grid".

Investors participating in the international forums have been refusing lately to fund the construction of coal-fired power plants polluting the environment. In the context of major accidents at nuclear power plants that had negative consequences, the world community is against the construction and operation of nuclear power plants. These circumstances foster the revision of energy policy in many countries and its shift in favor of renewable energy. The identical trends are observed in Mongolia that apart from considerable traditional resources has also substantial renewable energy resources. The forecast coal reserves in this country account for 175 billion t, oil – 205 million t, uranium – 68 million t [11, 12]. According to the assessment of the National Renewable Energy Laboratory of the USA, Mongolia is rich in wind energy resources, their forecast power is 1100 GW (only 10% of the country's territory have specific power above 600 W/h) [2]. According to the report of

Energy Charter issued in 2014, the anticipated power based on the use of solar energy can make up 2500 GW [2, 9, 10]. Graphical representation of this potential is demonstrated in Fig. 3.

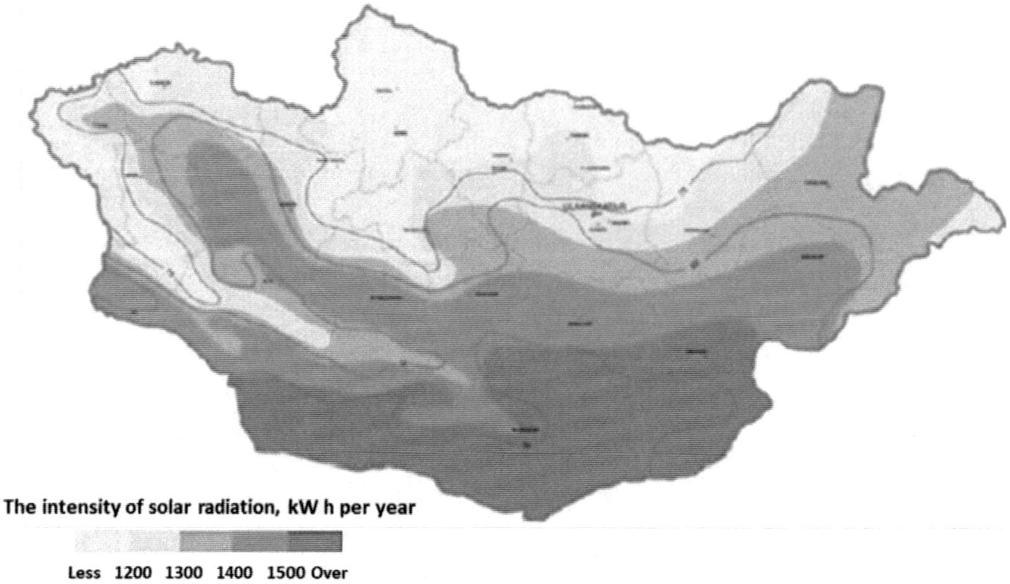

a

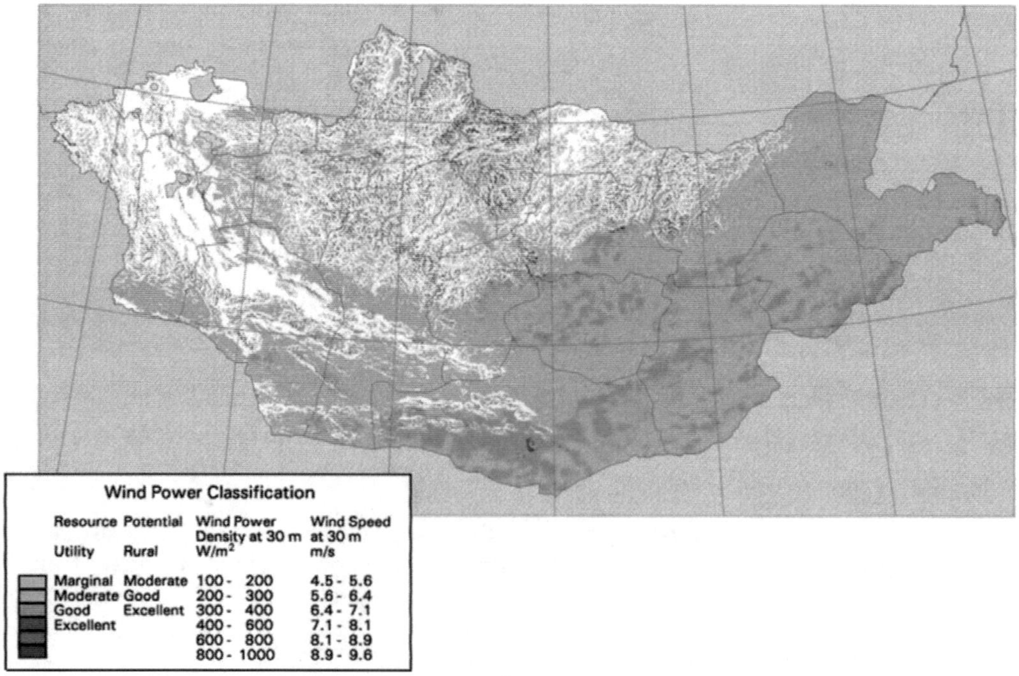

b

Fig. 3. Average annual solar(a) and wind (b) resources in the planned solar-wind system "Gobitech".

Despite substantial energy resources of various kinds (conventional and unconventional), there is power shortage which has been growing lately along with a permanent annual increase (by 7-10%) in electricity consumption [13]. Unfortunately, no measures are taken to eliminate the shortage and cope with this situation, unfavorable for the economy. For example in 2016 electricity import from the neighboring countries reached 1420 million kWh, its main part makes up the difference of the two-way power flow by the interstate transmission line Mongolia – Russia. The established energy cooperation is normally focused on electricity import. However, at the same time it is obvious that such a cooperation principle is not optimal to integrate the power systems of neighboring countries, and to create and develop the interstate power interconnection in Northeast Asia.

3 Specific features of the interstate power interconnection and new opportunities for Mongolia

The interstate power interconnection in Northeast Asia (Asian Super Grid) is still in the initial stage. Currently, there are only individual transboundary transmission lines mainly with a voltage of 110-220 kV with a transfer capability of 100-150 MW. These transboundary lines include the existing lines between Russia and Mongolia, Russia and China, China and North Korea, China and Mongolia [6-8]. Since 2011 an export 500 kV transmission line connecting the Amur substation (RF) and the Heihe substation (China) with a transfer capability of 750 MW and a DC link has been in operation [14, 15].

The investigations into the interstate power interconnections involve research institutes of Russia, including those of the Russian Academy of Sciences, Republic of Korea, China, Japan, Mongolia and other countries, and Asia-Pacific Energy Research Center in Tokyo (Japan). A generalized characteristic of power industries of potential participants in the large-scale project "Asian Super Grid" is presented in Table 1.

Table 1. A generalized characteristic of power industry of Mongolia and the NEA countries.

Index	China (2013)	Russia		Republic of Korea (2013)	Japan (2012)	Mongolia (2012)	DPRK (2012)
		Siberia* (2013)	The Far East* (2013)				
Area of the territory, million km^2	9598	5115	6169	99	373	1565	121
Population, million people	1357.4	19.3	6.3	50.0	127.6	2.8	24.8
Power consumption, TWh	5322.3	205.3	31.61	474.9	991.6	5.2	1.2
Power consumption per capita, kWh/person	3921	10637	5017	9498	7771	1857	734
Power generation, TWh	5347.4	197.4	35.2	517.1	1094.0	5.2	21.5
Installed capacity of power plants, GW	1247.4	49.3	9.1	91.0	287.3	1.02	7.22
including Thermal power plants:	796.4	25.0	5.7	56.3	188.9	0.97	2.96
coal-fired	758.1	24.2	5.2	24.5	50.9	0.88	2.76
gas and heavy oil –fired	38.3	0.8	0.5	31.8	138.0	0.09	0.2
Hydropower plants	280.0	24.3	3.3	6.5	48.9	0.03	4.26
Nuclear power plants	14.6	-	-	20.7	46.1	-	-
Renewable energy sources	86.8	-	-	3.5	3.4	0.01	-

**Power generation, power consumption, and installed capacity are given for interconnected power systems of Siberia and the Far East, respectively.*

With the proper energy policy and scientifically grounded concept of Mongolia's energy development, this country can take an appropriate place in the energy space of Northeast Asia, and its role in the construction and expansion of the interstate power interconnection in Northeast Asia (Asian Super Grid) can become significant. Implementation of such a concept will open up the following possibilities at least for Mongolia:

1. The regions of Northeast Asia with a high demand for electricity due to shortage of energy resources will be provided with electricity.
2. The renewable power generation will become more uniform and its efficiency will rise.
3. Mongolia's electricity export and import will change qualitatively: the country will turn from an importing country into an exporting-importing country.

Solving these problems foregrounds the technical and technological issues related to the connection of the regions by interstate UHV 500, 750 kV AC and ±400, ±600, ±800 kV DC overhead and cable transmission lines.

4 Formation of electricity market

The above circumstances and the level of the country development generate the need to develop a national concept of Mongolia's participation in the large-scale project of Interstate power interconnection in Northeast Asia. To preserve and expand its energy market in the near future Mongolia should first of all solve the following problems:

1. Take measures to create favorable conditions to attract investors from interested countries, including those from Mongolia.
2. Based on Mongolia's energy development strategy, perform a feasibility study and then implement the projects for construction of new power sources and transmission lines jointly with stakeholders concerned. In the low-population areas of the desert, it is necessary to envisage the construction of solar and wind power plants with a capacity of above 4000 MW, and new transboundary overhead transmission lines.
3. Organize energy cooperation with the NEA countries; enhance the effectiveness of such integration–based cooperation, considering seasonal load differences in each country, difference in time zones and specific features of climatic conditions.

The creation of the power interconnection infrastructure requires first of all interest and support of neighboring and other countries, as well as a feasibility study on export-import power delivery.

According to the authors of [1, 4], Mongolia has a transboundary 220 kV overhead transmission line connected with the electric power systems of Russia and China. This makes it possible to gain the necessary experience in operation of such lines. In the event of the development and expansion of operating interstate electric ties involving huge renewable energy resources of the Gobi area and coal, Mongolia can make a great impact on the formation and expansion of the interstate electric ties in NEA.

Japan is second only to China in the level of electricity production and consumption in the region. At the same time, this country is the first in nuclear energy development in NEA. Due to the limited indigenous natural energy resources, and imported power to meet domestic demand, Japanese power industry should solve two problems: decrease reliance on power import and reduce unfavorable impact of thermal power industry on the environment. In these conditions, the expansion of interstate electric ties with neighboring countries and import of green electricity from Russia and China could considerably improve the environmental situation in the country and meet external electricity demand. Different projects are offered for transboundary transmission lines between Japan and these countries. These however require additional feasibility studies on the construction of overhead and cable transboundary transmission lines.

South Korea is characterized by a very high level of power industry development. This country is a leader in electricity production per capita compared to the other NEA countries (Table 2). The structure of generation capacities in South Korea is similar to that in Japan. Korea is highly interested in green power import from Russia and China. However, it encounters serious political and financial barriers.

Table 2 presents the indices of energy available in NEA countries, and the world electricity production and consumption compared to these indices for Mongolia.

The data are of interest in terms of specific electricity consumption (per capita) that varies greatly with respect to the average level.

Canada, the USA and South Korea from NEA countries stand out in terms of per capita electricity consumption. Mongolia power consumption index is below the average level. However, this country has favorable conditions for the improvement of this index owing to enormous reserves of traditional and renewable energy. Moreover, this allows Mongolia to participate in electricity market as a power exporter.

Table 2. Global power production and consumption.

No	Countries	Generation GWh	Population, thousand	Consumption per capita kWh/person
1	China	5649500	1376622	4103.89
2	USA	4297300	323394	13288.13
3	India	1208400	1288306	937.98
4	Russia	1064100	146545	7261.25
5	Japan	1061200	126980	8357.22
6	Germany	614000	81174	7564.00
7	Canada	615400	34850	17658.54
8	Brazil	582600	205738	2831.76
9	France	555700	64513	8613.77
10	Republic of Korea	517800	51431	10067.86
11	Mongolia	5541.7	3000	1847.23
12	Niue	3	14	214.29
Total in the world		23536500	7300000	3224.18

5 Prospects for the interstate power interconnection in Northeast Asia ("Asian Super Grid")

The studies [1, 3-8] show that the prospects for the power industry integration in Northeast Asia are determined by the projects for the development of interstate electric ties in the energy sector and large-scale projects on the interstate power interconnections. The projects differ in the form of cooperation: transboundary trade, power export, connection of national or local electric power systems of neighboring countries for joint (or parallel) operation; structure of generating capacities (relationship between power sources of various types); technical and technological solutions for transboundary transmission lines (Table 3).

Among the potential directions of interstate electric ties in NEA, that are presented in Table 3, of interest is transmission of power generated from a solar-wind system in the Gobi desert.

Table 3. Prospective interstate electric ties among NEA countries.

Interstate electric ties	Length, km	Voltage, kV	Transfer capability, GW	Transmitted power, TWh/year	Tentative cost, billion USD
Russia – China					
Bratsk – Ulan-Bator – Beijing	2250	±600	5-6	18	1,8
Bureya HPP – Harbin	700	±400	1,0	3	2,2
Large-scale power export project	3400*	±600	10*	60*	18*
Erkovetskaya CPP – Shenyang	1300	±600	3,6	20	8,8
DC system "Ust-Ilimsk – Khabarovsk"	5000	±750	10,0	40	16,5
Russia – Korean peninsula					
Vladivostok – Chongjin	370	±500	0,5	3	0,13
Vladivostok – Pyongyang – Seoul	1150	±500	4,0	7	4,8
South-Yakutian HPPs – Shenyang – Seoul	2400	±750	5,0	20	10,5
Russia – Japan					
Sakhalin – Hokkaido – Honshu	1850/1400**	±600	4/3	24	9,6
Sakhalin – Hokkaido	500/50**	±500/±400	4,0	24	6,7
Asian Super Grid					
Gobitec – Mongolia, Russia, China, Korea, Japan	7300	±800	100*	200*	56,7*

* *generalized project indices.*
** *in numerator– total length, in denominator – length of submarine cable.*

Russia proposes the projects for the hydropower plants oriented to long-distance power transmission. One of them is a 9050 MW hydropower plant on the Lena river, currently under design. Average annual output makes up 7.8 TWh, the hydropower plant construction cost is about 3.6 billion USD. It is also planned to construct Mokskaya HPP on the Vitim river in the Republic of Buryatia. Power generated by this plant will be supplied to deficient regions in the Far East, Mongolia and China. The Bureya and Zeya HPPs operating in the Amur power system can also be considered export-oriented. Implementation of these projects will enable power supply to consumers in Siberia and the Far East. Moreover, they will undoubtedly contribute to the increase in the maneuverability of the transboundary power transmission, and enhancement of the efficiency of transboundary power systems.

Involvement of Mongolia's power system in the project "Bratsk – Ulan-Bator – Beijing" aimed at exporting surplus power from the Siberian power system can enhance the efficiency of this project, and create good conditions for power export to neighboring countries. In 2005, JSC "RAO UPS of Russia" and State Grid Corporation of China concluded a contract on mutually beneficial energy cooperation. According to this contract, Russia should export 60 trillion kWh to China annually [13]. The first stage of this project ended with connection of the Far Eastern electric power system to the power system of Heihe (Northeast Province of China). This enables the interconnected power system of the East to supply up to 3.5 billion kWh of power yearly to the northeastern provinces of China.

The most effective project among these is the interconnection of the regional power system of the Far East with the power system of the Republic of Korea through the transboundary transmission line "Vladivostok – Pyongyang – Seoul", 1150 km long. The expected economic benefit, i.e. saving of investment, from interconnecting the capacities and using the difference in time zones (9-11 hours) reaches 6-7 billion USD [1].

Another interesting project of transboundary power transmission is the project "Sakhalin – Hokkaido – Honshu", the so called "power bridge". The idea of this bridge appeared in the 1990s. This project suggests the construction of two export-oriented condensing power plants (4 GW Solntsevskaya coal-fired power plant - in the first stage, and Vakhrushevskaya combined cycle power plant - in the second stage), and ±400 kV DC transmission line "Sakhalin – Sapporo – Tokyo", 1600 km long, with two submarine cables at the bottom of the La Perouse Strait (50 km) and Sangarsky strait (40 km).

Participation in the construction and development of the NEA interstate power interconnection (Asian Super Grid) requires that Mongolia perform a feasibility study on the construction of the wind -solar system in the Gobi desert (Gobitec) to export cheap power to China, Republic of Korea, Japan, and Russia, and further implement the project. In this context it will be first of all necessary to solve the issue of constructing a (±600 kV) DC transmission line and a (500 kV) AC transmission line.

The creation of the 100 GW wind-solar system in the Gobi desert will give an impetus to the construction of Asian Super Grid and creation of several interconnections: A Mongolian ring: Gobitec – Mugden – Harbin – Kharanuur; the Japanese Sea ring: Seoul - Pyongyang – Hokkaido – Honshu; and a big ring: Bratsk – Urgalsk – Sakhalin – Tokyo – Shanghai – Beijing – Ulan-Bator – Irkutsk (Fig.1). This project will obviously have a long-term implementation period but it will have a tremendous benefit, for example, a reduction in the fuel costs alone will make up 10.0 trillion USD a year.

6 Conclusion

The interstate electric power interconnections start with the construction of transboundary transmission lines ensuring joint operation of national or local power systems of two, less often three countries. Such transmission lines tend to operate in an alternate mode, and this allows the countries involved to enjoy the synergetic effect. A comparative analysis of the expansion and operation costs of electric power systems of neighboring countries for the cases of their integration and their independent operation shows that such electric ties can be rather effective. This can be exemplified by the interstate power bridges: "Bratsk - Ulan-Bator - Beijing"; "Vladivostok - Pyongyang- Seoul"; Shivee - Ovoo- Kharanuur - Sakhalin"; Shivee – Ovoo - Shenyang - Pyongyang - Seoul" and other transboundary transmission lines connecting power systems of different countries with non-coincident seasonal peak loads. Active participation of Mongolia in the interstate power interconnection in NEA requires that Shivee- Ovoo energy facility and Gobitec be connected to the electric network covering the Russian Far East; Northern and Northeastern regions of China; Japan; South and North Korea.

To summarize, we can conclude that there are a lot of proposals for the interstate power interconnection of electric power systems. However, they are not coordinated with one another, therefore the parties concerned, involved in the NEA interstate power interconnection should carry out joint scientifically grounded studies to solve the energy issues existing in these countries.

References

1. Yu.N. Kucherov, O.M. Kucherova, L. Kapoji, Yu.N. Rudenko, *Reliability and effectiveness of large-scale trans-national electric power systems* (Novosibirsk, Nauka. 1996) (In Russian)
2. *In-depth review of the investment climate and market structure in the energy sector of Mongolia* (Belgium, Energy Charter Secretariat, 105, 2013) http://www.energycharter.org/fileadmin/DocumentsMedia/ICMS/ICMS-Mongolia_2013_ru.pdf
3. H. Bahar, J. Sauvage, OECD Trade and Environment Working Papers, **2**, 78 (2013)
4. *Eastern vector of Russia's energy strategy. Current state, look into the future* (Novosibirsk, Geo. 2011) (In Russian)
5. S. Mano, B. Ovgor, Z. Samadov etc, *Gobitec and Asian Super Grid for Renewable Energies in Northeast Asia* (Belgium, Energy Charter Secretariat. 2014)
6. *Wind Energy Resource Atlas of Mongolia, National Renewable Energy Laboratory* (NREL/TP-500-28972., US. 2001)
   https://digital.library.unt.edu/ark:/67531/metadc721321/
7. J.Y. Yoon, D.W. Park, H.Y. Kim, *Proceeding of 6th International Conference – Asian Energy Cooperation:Forecast and Realities* (Irkutsk, 2008) (In Russian)
8. W. Chandler, Ch. Shiping, H. Gwin, L. Ruosida, W. Yanjia, WWF Report, **2**, 71 (2014)
9. *Mongolia's Climate Reference Book* (Ulan-Bator, GMS. 1984)
10. *Climate of Mongolia* (Ulan-Bator, Gosizdat. 1985)
11. *Brief data on coal, oil and oil shale deposits of Mongolia, Mineral Resources and Petroleum Authority of Mongolia* (Ulan-Bator, Geological Information Center. 2007)
12. *Uranium resources, production of Mongolia (RED BOOK) Uranium, 2009: Resources, Production and Demand* (OECD, IAEA. 2010)
13. *Statistics of energy sector 2016* (Ulan-bator, Coordination Committee for the Energy Sector, 2017)
14. S.V. Podkovalnikov, V.A. Saveliev, L.YU. Chudinova, External Economic Relations. **4,** 118-130 (2015) (In Russian)
15. *Interconnection in APEC Region. Current Status & Future Potential* (Tokyo, Asia Pacific Energy Research Center, 2000)

Data processing and optimization system to study prospective interstate power interconnections

Sergei Podkovalnikov[1,*], *Ivan* Trofimov[1], and *Leonid* Trofimov[1]

[1]Melentiev Energy Systems Institute Siberian branch of the Russian academy of sciences, Department of Electric Power Systems. 664033, Irkutsk, Lermontov St., 130. Russia

Abstract. The paper presents Data processing and optimization system for studying and making rational decisions on the formation of interstate electric power interconnections, with aim to increasing effectiveness of their functioning and expansion. The technologies for building and integrating a Data processing and optimization system including an object-oriented database and a predictive mathematical model for optimizing the expansion of electric power systems ORIRES, are described. The technology of collection and pre-processing of non-structured data collected from various sources and its loading to the object-oriented database, as well as processing and presentation of information in the GIS system are described. One of the approaches of graphical visualization of the results of optimization model is considered on the example of calculating the option for expansion of the South Korean electric power grid.

Introduction

The relevance of research in the field of expansion of interstate power interconnections is determined by the expansion of globalization processes in the world. To meet electricity demands, there is a need to integrate electricity systems both within the country and among countries. Such interconnections are already functioning in Europe, North and South America and other regions of the world [1]. However, in order to justify the creation of such interconnections, qualitative preliminary calculations are required to predict the effectiveness, ensure the reliability of the interconnected power systems, and much more.

Conducting scientific research in this area requires a lot of preliminary work on the collection and analysis of a huge amount of information, complex technical and economic calculations, significant intellectual and information resources. Without the presence of advanced information technologies, specialized in this subject area, research takes a huge amount of time and money. Multidimensional results of calculations are also difficult to imagine without a specialized graphical interface for data representation and processing.

The Data Processing and Optimization System (DPOS), is being developed by the authors, integrates in itself the technologies for collecting and processing non-structured data from open sources; the technology of information restructuring in the object view; the

* Corresponding author: spodkovalnikov@isem.irk.ru

technology of graphical and cartographic data representation in the GIS system; and the mathematical model for optimizing the expansion and operating modes of the power systems/interconnections – ORIRES. As a result of the DPOS usage, the efficiency and quality of scientific research is increased in the field of expansion of interstate electric power systems being conducted at the Melentiev Energy Systems Institute of the Siberian Branch of the Russian Academy of Sciences (ESI SB RAS).

1 The data processing and optimization system architecture

In ESI SB RAS conducted is research into the possibilities of combining national power systems of various countries (mainly in the Northeast Asia region) into interstate power interconnections. The authors are developing the DPOS for complex information and analytical support of scientific research in the field of expansion of interstate energy interconnections. At calculations the model for optimizing expansion and modes of electric power systems ORIRES, integrated in DPOS is used.

The DPOS is based on an object-oriented database (OODB) [2] to store and processing of information. All information in the OODB is structured as objects. Types of objects in the OODB are independent and different - power plant, city, region, country, power interconnection, power line and others. Each object is a record containing a unique object identifier, a set of parameters describing it, and the values of each parameter in text and numeric form, which can be stored by year.

A user interface for inputting / outputting information from OODB has been developed, using graphical and cartographic representation of energy/power data. For the presentation of materials, combined fragments of GIS maps with the main components of the electric power infrastructure are used. The information can be presented in retrospect from the 90s to the present, and also taking into account the expected medium and long-term outlook.

An interface is also being developed to work with the optimization model ORIRES integrated into DPOS to perform complex optimization studies of various scenarios for the formation of interstate power interconnections, including studies of the prospects for the creation of the Asian Energy Supergrid [3-4].

The main structural-functional blocks of DPOS, Fig. 1:

• Procedures (technologies) for collecting information resources,
• Object-oriented database,
• Interface for working with the mathematical model ORIRES,
• Data representation interface in the integrated GIS system,
• Analytical Internet Service for presenting results to the Internet [5-6].

Functional stages of DPOS operation:

1. Collection of various untyped data from open sources and Internet resources;
2. Loading information into an object-oriented database. Object structuring of collected data in OODB;
3. Presentation of information in the interface for generating and adjusting the parameters of mathematical models for the GAMS optimizer;
4. Launch of the integrated in DPOS optimizer GAMS, obtaining variants of mat-model solutions;
5. Use ready-made solutions in a table view or upload them back to the OODB for further processing;
6. Presentation of information in the interface of the GIS-system for building interactive maps, graphs and diagrams, with the possibility of retrospective and prospective data analysis;
7. Export of processed information to an external Analytical Internet Service in order to attracting interest of the scientific community to the results of research.

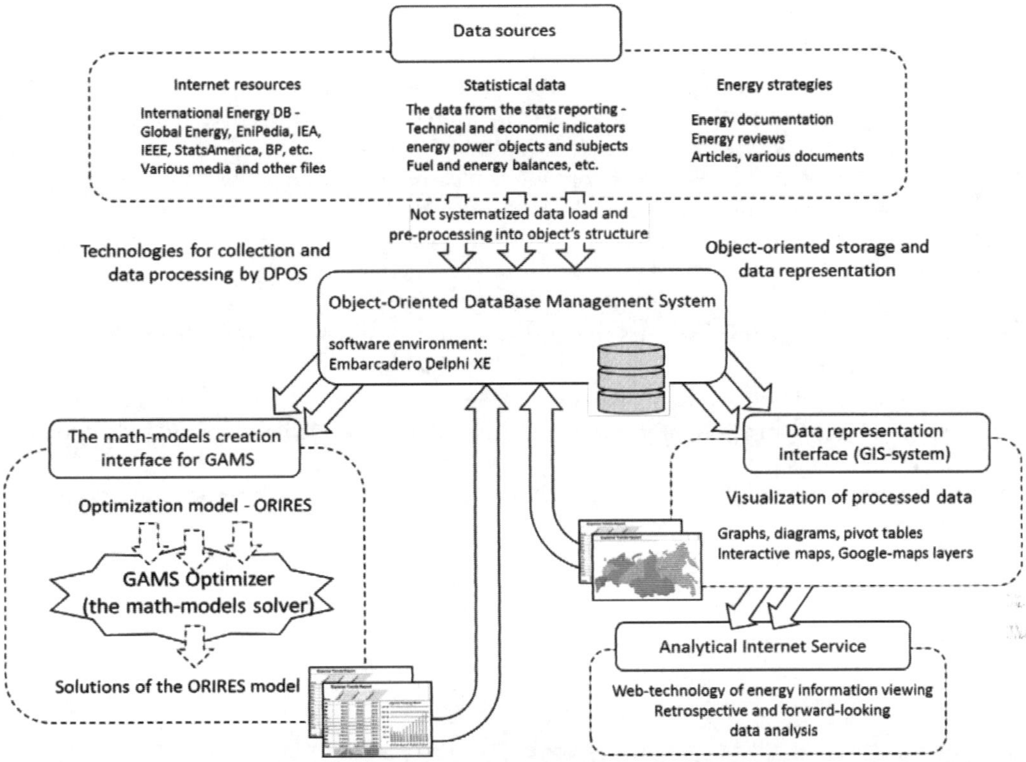

Fig. 1. Structural-functional scheme of the DPOS with integrated GIS-system and model ORIRES.

When implementing DPOS, modern mathematical, information and web technologies are used:

• The Embarcadero Delphi XE software environment, Google Maps API functions, SHP and XML files containing geographic and other information about objects of study are used as a basis for constructing the GIS-system and the Object-Oriented Database.

• To create a web application, modern web technologies based on PHP, Ajax, JQuery, JavaScript, MySQL, Bootstrap are applied.

• The forecast optimization model for the expansion of electric power systems ORIRES is written in the form of a set of mathematical formulas and equations solved through a specialized GAMS solver, the initial data for it, as well as the results of the solution can be obtained / loaded into an OODB for their subsequent processing and presentation.

2 Storage and data processing in the Object-Oriented Database

Official Internet sources of information about energy/power facilities and subjects of different countries have limited or paid access [7]. Own databases are not available for general use, and public information is mostly poorly structured for automatic machining.

To conduct qualitative research it is necessary to have the maximum completeness of information about the objects or processes being studied. In particular, in order to determine the initial state of the Interconnected Power System (IPS), it is necessary to have information about the installed and planned electricity generating capacities of this IPS. When collecting and obtaining such information, the researcher faces subjective and objective difficulties. Often sources of information are scattered and located on various Internet resources.

The authors of this work created a technology for collecting non-structured data and putting it in an object structure with subsequent verification and analysis.

Feature of the technology for data collection and processing is the algorithm developed by the authors for obtaining data stored in the form of ontologies on specialized open sources. With the help of this algorithm, information about practically all energy/power objects in the world can be recorded in the Object-Oriented Database and then transferred to the object structure [8].

As a result, a thematic OODB is formed with a unique information gathered within a single structure. Its use greatly expands the possibilities and improves the quality of scientific research in this subject area.

For the DPOS programming interface, original procedures and algorithms for uploading data are developed, allowing users to work with information from OODB, analyze it, edit it and conduct various calculations. The calculation results can also be uploaded to the OODB for their subsequent use.

Objects in the OODB are represented as records with a set of parameters and values of the studied objects of different types, and have auxiliary media files that supplement information on objects: photos, articles or links to objects on the Internet. Objects in OODB can be various objects of the studied subject area: power plant, city, region, country, power interconnection, power line and others, Fig. 2.

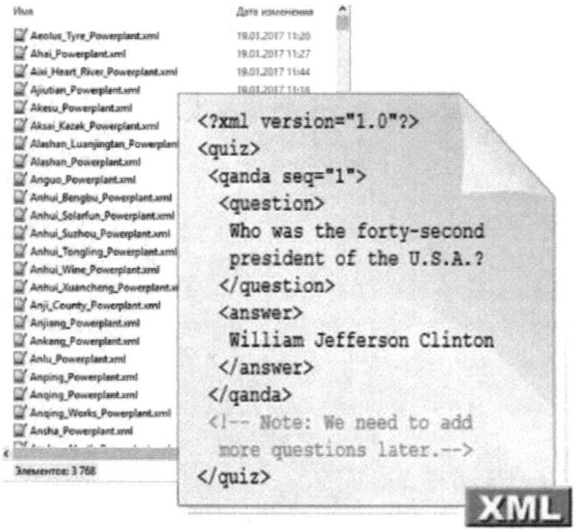

Fig. 2. An example of the content of object information in OODB in XML format.

Statistical information processing from OODB is carried out using a special query language that allows retrieving the parameters of objects and performing their sorting, sampling of

Objects have an extensible set of parameters that can be added while working with OODB. In particular, for the power plants, the parameters of their cartographic binding and the possibility of representation on maps in the GIS-system are added, Fig. 3. The shaded areas in the figure are the territories served by IPS of North China and IPS of Northeast China.

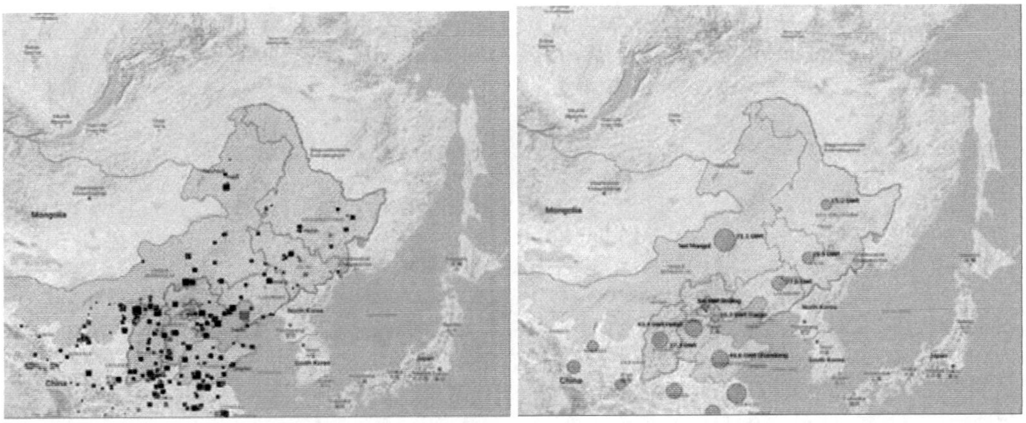

Fig. 3. Installed capacity of the North China IPS and North-East China IPS.

The collected information can be aggregated across different territorial and technological hierarchical levels [9] (country, IPS, etc.), which opens up opportunities for scaling the model - the creation of different levels of detail in the calculation scheme with different number of nodes and electrical connections. In Fig. 4 there is an example of a diagram for the Interstate Power Interconnection in the Northeast Asia region.

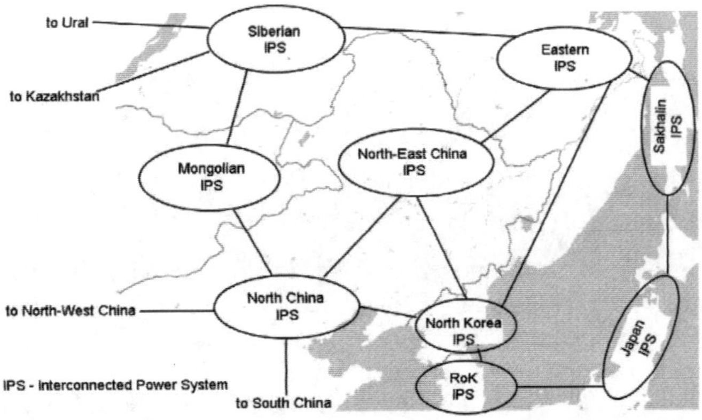

Fig. 4. The diagram of the Interstate Power Interconnection of Northeast Asia.

The primary list of objects in the OODB is formed from open sources on the Internet – from EnergyPedia.org etc. [10]. IPS as well as individual power plants are represented in the OODB in the form of objects and their parameters. A number of other parameters have been introduced for this type of objects - daily load levels and graphs, aggregated volumes of power generating capacities by type of plants and so on. All OODB objects have numerical parameters and reference information, i.e. text parameters: name, province and others. The values of these parameters are used as input data integrated into the DPOS optimization model of ORIRES.

3 The optimization model's solutions visualization in the DPOS interface

The tasks of the ORIRES model include the calculation of various options for the formation of the IPS, the calculation of the optimal expansion of IPS, take into account capital investments, fuel costs, annual operating costs, variables that characterize intersystem flows, etc. [11]. The model solutions on the output from the mathematical optimizer GAMS are presented in a large amount of information on many pages, Fig. 5.

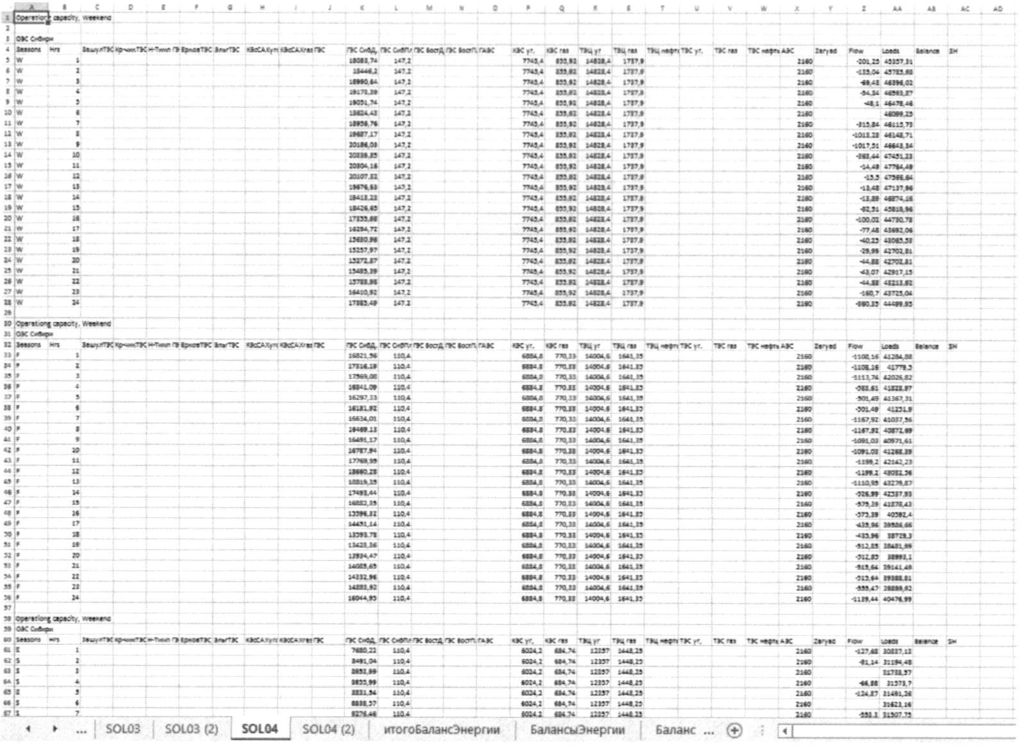

Fig. 5. Output data example of GAMS optimizer solution.

Multidimensional results of calculations are difficult to observe without a specialized graphical interface. To perceive this amount of information by an expert and his further analysis, specialized procedures are required to aggregate the results of the solution into different types of summary tables and diagrams.

The authors developed a special approach to step-by-step presentation of the results of the model solution. In the first step, the volumetric matrices of the primary detailed output data are converted into more aggregated tables of calculation results, which, while still sufficiently detailed, are convenient for analysis by experts. This form of visualization of the result of the optimization calculation is presented in the form of the annual energy and economic balance of South Korea's IPS, Table 1. In the balance sheet, exports are presented as a negative number, imports as positive. This table is only a small fragment of the presentation of the results of calculations in step 1.

Table 1. South Korea, Annual energy and economic balance, 2035.

Capacity type	Output, GW-hour	Input, GW-hour	Total, GW-hour	Optimum power, MW	Fuel cost, mln.$	Service cost, mln.$	Total cost, mln.$
Nuclear		325968	325968	44000	2607	6504	9112
Cogeneration coal		7119	7119	1100	256		256
Thermal coal		340408	340408	57100	12935	2384	15319
Thermal oil		3774	3774	1000	868		868

Cogeneration gas		46402	46402	11000	6403		6403
Thermal gas		107927	107927	33594	15109		15109
Hydro		14117	14117	1900		27	27
Renewable		2976	2976	4700			
Total generation		848695	848695	154394	38181	8916	47097
Storage	-3721		-3721				
Total	**-3721**	**848695**	**844974**	**154394**	**38181**	**8916**	**47097**

In step 2, there is further aggregation and, most importantly, visualization of the output data in the form of graphs and diagrams, which is convenient for presenting results in reports and presentations. As an example, we consider a variant of plotting the graphs reflecting the hourly daily load in South Korea's IPS, meeting it by various types of generating capacities (nuclear, coal, gas, hydraulic, wind, etc.), as shown on the right-hand side of Fig. 6. The left part of Fig. 6 shows the columns representing the working and installed capacities, as well as limits on their expansion in the considered period (until 2035) for IPS of South Korea:

- Column 1 reflects the upper limit of how much capacity of power plants in a node can be increased compared to the current level.
- Column 2 reflects the "optimal" solution obtained as a result of the calculation by the ORIRES model, which meets the growing demand with minimal cost.
- Column 3 reflects the current mix of installed capacity.
- Columns 4 and 5 - the limits of the operating range in which the power output of each type of power plant can vary, which depend on their technical characteristics.

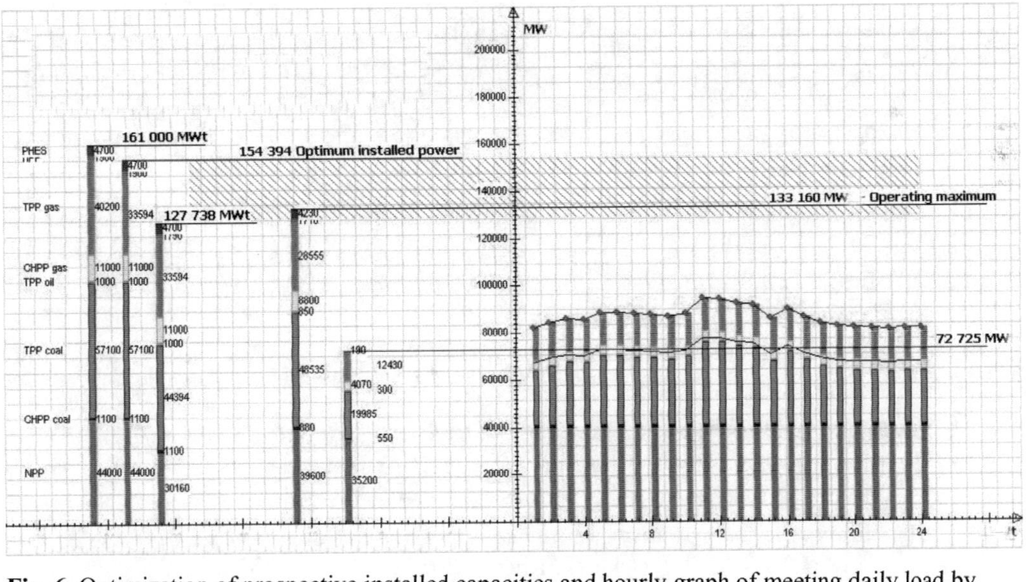

Fig. 6. Optimization of prospective installed capacities and hourly graph of meeting daily load by various types of power plants.

The diagram under the daily load graph shows how much power and what type of power plant was used for each hour. Usually considered are eight different types of daily load graphs (summer weekdays / holidays, autumn weekdays / holidays, winter weekdays / holidays, spring weekdays / holidays). Multiplying and totaling each of the eight daily generation of electricity by the corresponding number of days in a year, the amount of electricity produced per year is obtained. At the same time, the amount of electricity generated per year will be broken down into generation by types of power plants. In addition, the model calculates the cost of fuel, operational and investment cost of each type of power plant. Using it, we can get the final cost of generated electricity for the year.

The proposed approach to visualizing the results of the solution of the model increases the efficiency and convenience of perceiving a large amount of information aggregated and depicted in generalized graphs.

Conclusion

The authors developed the basic technologies for the construction of a specialized DPOS, including the procedures for collecting, storing and processing of poorly structured data, an interface for flexibly adjusting the parameters of the mathematical model ORIRES and its solutions by the GAMS optimizer, the presentation interface for retrospective and prospective analysis in the GIS system.

A new form of information presentation is proposed for more convenient comparison of different solutions of the optimization model using graphical visualization.

The DPOS proposed by the authors allows significantly increasing the efficiency of scientific research in the field of expansion of interstate power systems and improving the results of the solution qualitatively by using a convenient graphical interface with the ability to configure input and output parameters for the optimization model.

References

1. N.I. Voropai, S.V. Podkovalnikov, V.A. Savelyev, L.Yu. Chudinova, Electricity. Transmission and distribution, **3** (42), 8 (2017)
2. Douglas K Barry. *Web Services, Service-Oriented Architectures, and Cloud Computing. Database Articles. Object-Oriented Databases* (2013)
3. S.V. Podkovalnikov, V.A. Saveliev, L.Yu. Chudinova, Studies on Russian Economic Development, **26** (4), 403 (2015)
4. D. Gao, O. Khamisov, D. Sidorov, International Journal of Artificial Intelligence, **13**, 120 (2015)
5. I.L. Trofimov, *Energy Statistical Analytical Service. ESAS. World Energy on an interactive map* (2015)
6. L.N. Trofimov, I.L. Trofimov, *Energy Industry in Russia in 21 Century: Innovative development and management,* 170 (Irkutsk, Russia, 2015)
7. *Global Energy Observatory. Power Plants* (2017)
8. I.L. Trofimov, *Zbornik radova konferencije MIT 2013,* 706 (Vrnjacka Banja, Serbia and Budva, Montenegro, 2014)
9. I.L. Trofimov, *System research in the energy sector. Collection of works of young scientists ISEM SB RAS,* 215 (Irkutsk, Russia, 2013)
10. *Energy Pedia. Portal: Power Plants* (2017)
11. L.S. Beliayev, S.V. Podkoval'nikov, V.A. Savel'yev, and L.Yu. Chudinova, *Efficiency of Interstate Electric Ties* (2008)

A shift in the paradigm of energy cooperation between Russia and Northeast Asia countries facing new global and regional challenges: from predominant sale of energy resources to innovation and technology cooperation

Boris Saneev*, *Alexander* Sokolov, *Anatoly* Lagerev, *Sergei* Popov, *Irina* Ivanova, *Alexander* Izbuldin, *Anatoly* Korneyev, *Svetlana* Muzychuk, *and Dmitry* Sokolov

Melentiev Energy Systems Institute of Siberian Branch of the Russian Academy of Sciences, 130, Lermontov str., Irkutsk, 664033, Russia

Abstract. The paper is concerned with the specific features of Russia's economic development in a new economic environment that caused the need to revise the priorities of energy policy. The research presents the initial conditions, targets and strategic directions of energy development in the East of the country. The focus is made on the priority lines of innovation and technology cooperation between Russia and Northeast Asian countries in the field of energy, and recommendations on necessary conditions and initiatives for their successful implementation are given.
Key words: East Siberia, the Far East, Eastern energy policy, energy sector, energy cooperation interstate innovative-technological cooperation in the energy field, mechanisms of energy strategy implementation.

1 Eastern vector – priority line in the energy development of Russia in the first half of the 21st century (main concepts)

Specific features of Russian economic development in a new economic environment caused the need to revise earlier adopted priorities in the economic and energy policy of the country.

The national interests of Russia require intensification of mutually beneficial energy cooperation between Russia and China, Japan, Korea and other countries of East and Northeast Asia[1].

In the official documents, this priority direction in the energy development of the country was named "Eastern Vector of Russia's energy policy" [1, 2]. The idea of this direction can be briefly formulated as follows:

* Corresponding author: saneev@isem.irk.ru

[1] According to the publications and some official documents of Japan, China and the Republic of Korea, the Northeast Asian countries include Mongolia, China, North Korea, South Korea, and Japan

- creation of new energy centers in the East of the country[2] will increase energy security of Russia, restore and strengthen energy ties among the regions, solve many crucial federal, interregional and regional problems;
- creation of a wide energy infrastructure including interstate gas and oil pipelines, and transmission lines in the east of Russia and Northeast Asia will reduce the cost of energy, improve reliability of energy supply to consumers in different countries, and help solve the environmental problems.

Eastern energy policy of Russia being part of the economic policy is not an end in itself but a tool for solving many principally important federal, interregional and regional problems;

1. Socioeconomic and geopolitical objectives:

– *social* – to increase comfort and improve living conditions of the population in the eastern regions of Russia.

– *economic* – to enhance the efficiency of the economy of Russia's East and its competitiveness, increase the availability of resources and accessibility of territories, expand active economic space of Russia, create conditions for attraction of foreign capital and advanced technologies;

– *political – to* consolidate and integrate entities of Federation, strengthen the unity of economic and energy space of the country;

– *geopolitical – to* strengthen Russia's position in the world economic system and in the community of states in Central and Northeast Asia.

2. Energy objectives:

– to increase energy security of the country and regions;

– to increase adaptability and reliability of energy and fuel supply to consumers;

– to improve the territorial and production structure of the national energy sector, especially in the eastern regions of the country;

– to build the energy transport infrastructure in Russia's East (systems of oil and gas pipelines, and power transmission lines) and create a single energy transport space in Russia.

Eastern regions in the country – East Siberia and the Far East – with their great economic and energy potential are of paramount importance for the implementation of Russia's national interests in this strategically important region of the world.

2 Initial conditions, targets and priority directions in the energy development of Russia's East

East Siberia and the Far East of the Russian Federation possess a large fuel and energy potential:

The hydro power potential is estimated at 81 % of that of the country. In East Siberia and the Far East there is 35 % of coal, 15 % of oil and 12 % of natural gas proved reserves.

These regions play an important role in the production of energy resources in Russia: *in 2015 the eastern regions produced 18,5 % of electricity generated in the country, extracted 34,7 % of coal, 12,1% of oil, 7,2 % of natural gas and refined 10,5 % of crude oil.*

Russia developed a large number of policy documents determining the strategic development of the economy and energy in the East of the country until 2030 in the context of energy cooperation between Russia and Northeast Asian countries. These documents are: "Energy Strategy of Russia until 2030" [2], "Program for Creation of Unified Gas Production, Transportation and Supply System in East Siberia and the Far East with Potential Gas Export to the Markets of China and other APR Countries" (Eastern Gas Program) [3], "Strategy of

[2] In the energy strategies of Russia, the East of the country includes the regions of East Siberia and the Far East

Socioeconomic Development of the Far East and the Baikal region until 2025" [4], "Strategy of Socioeconomic Development of Siberia until 2020" [5], and others.

Melentiev Energy Systems Institute SB RAS made a certain contribution to the development of the Eastern energy strategy of Russia. In 2007-2014, in the framework of feasibility studies on the Energy Strategy of Russia 2030, and research ordered by the regional authorities, the Institute conducted a great amount of studies on feasibility of long-term directions of energy development of the country and its eastern regions and some entities of the Russian Federation in their territories [6-11].

The strategy of long-term energy development of East Siberia and the Far East was developed considering some initial conditions and targets, of which the most important are the following two:

1. In the coming 15-20 years, Russia will be unable to develop the untouched territories of East Siberia and the Far East. Therefore, it is necessary to preserve and strengthen the base for the future development.

This can be done through the so called strategic scenario of economic development of these regions.

The strategic scenario of the economic development of the eastern regions suggests that the economy of Russia and its eastern regions should quantitatively and qualitatively near the current average European level. In doing so, the rates of the economic growth in the considered regions should be higher than on average in Russia. An increase in the share of these regions in the total population number of the country is also planned.

2. Energy development in East Siberia and the Far East in the considered time horizon will be aimed at both meeting the domestic demand for energy carriers and exporting Russian energy resources to the energy markets of NEA countries.

An analysis of the situation in the energy markets of China, Korea, Japan and other countries of Northeast and Eastern Asia shows that there a niche for the Russian energy resources in these countries. Russia is ready to provide these countries on mutually acceptable terms with the following energy resources from the eastern regions [2, 6, 12, 13, estimates by authors]: oil – from 69 million t in 2016 to 75-80 million t in 2030; natural gas – from 15 billion m^3 in 2016 to 50- 60 billion m^3 in 2030; coal – from 70 million t in 2016 to 75-85 million t in 2030; electricity – from 3.6 billion kWh in 2016 to 45-50 billion kWh in 2030.

Current state and prospects for energy development in the East of Russia (for two scenarios) are given in Table 1.

The energy development in East Siberia and the Far East up to years 2030-2035 requires huge investment: the cost for some of the forecast strategies of energy development in the east of Russia is estimated [6] at $200-250 billion only for new energy enterprises, including $80-85 billion to be invested in the development of oil and gas production, and main oil and gas pipelines.

Such a strategy of energy development in East Siberia and the Far East is unlikely to be implemented without attraction of foreign investment, which supposes close energy cooperation between Russia and NEA countries.

The Russian government, regional authorities and companies have started a large-scale development of energy resources in the East of the country.

At present, the material resources of the eastern vector of Russia's energy policy are based on several large fuel and energy projects aimed at markets of NEA countries:

First, construction of the oil pipeline "East Siberia – Pacific Ocean" with a capacity of 80 million t per year with an oil pipeline branch to Skovorodino towards China with a capacity up to 15-20 million t of oil per year.

Table 1. Production of energy resources in the East of the Russian Federation

Index	2015 (actual)	Forecast	
		2025	2035
Energy production , million tce, total	246	360-405	420-475
including:			
- East Siberia	151	190-215	230-260
- Far East	95	170-190	190-215
Oil, million ton, total	63	85-95	100-110
- East Siberia	37	45-50	55-60
- Far East	26	40-45	45-50
Natural gas, billion cubic meter, total	30	95-115	115-135
- East Siberia	8	35-45	50-60
- Far East	22	60-70	65-75
Coal, million ton, total	136	153-165	170-190
- East Siberia	96	90-100	105-115
- Far East	40	63-65	65-75
Hydropower, TWh, total	106	130-136	151-157
- East Siberia	91	110-112	118-122
- Far East	15	20-24	33-35

Note: the left-hand value of the interval corresponds to an index for the conservative energy development scenario, and the right-hand value – for the optimistic scenario
Source: forecast indices – estimates by authors

Second, "Gazprom" drew up a program to develop natural gas resources in the East of Russia, convert Russian regions to gas, and deliver natural gas to NEA countries.

Third, Russian electricity companies are conducting a feasibility study on large-scale electricity supply to China, Japan and other NEA countries.

Third, Russian electricity companies are conducting a feasibility study on large-scale electricity supply to China, Japan and other NEA countries.

For Russia facing the new global and regional challenges, it is principally important not only to cooperate with NEA countries and trade in energy resources, but also to pursue an active technology and innovation policy.

3 Priority directions of technology and innovation cooperation between Russia and NEA countries in the field of energy

3.1 Participation in the construction of new Russian oil and gas-chemical clusters and their joint control.

Oil and gas cooperation between Russia and NEA countries should take into consideration the following factors.

1. Currently, Russia at all levels is well aware of the need to deliver not only hydrocarbon raw materials to the international markets but also the products of advanced processing, with high value added. To this end, it is proposed to increase the production of oil products in the eastern regions of Russia, create new industries (oil-gas chemical, for example), since there is a high demand for their products in Russia and Northeast Asian countries. In the future Russia can become the major exporter of polymers in the region.

2. Oil and gas fields of the Siberian platform are unique in helium and ethane content. Natural gas from the fields of the Siberian platform contains up to 0.3-0.5 % of helium and

4.6-7.2 % of ethane. Helium reserves (category C_1+C_2) from the oil and gas fields of the Siberian platform are estimated at 8.5 billion m^3, or about 30 % of the world reserves.

In the future, Russia can become the largest helium exporter.

Ministry of Energy of Russia devised a plan for the oil and gas chemistry development in Russia until 2030. This plan was approved by the Governmental Commission on the Fuel and Energy Complex on 28.12.2011, and its implementation (stage II) [14] was approved by the Order of Ministry of Energy of Russia No.79 of March 01, 2012.

The plan envisages the creation of 6 new oil and gas chemical clusters in Russia by 2030:

– Northwestern
– Caspian
– Volga
– West-Siberian
– East-Siberian
– Far-Eastern.

Russia is interested in mutually beneficial cooperation with foreign companies to establish new oil and gas chemical clusters in the East of the country, and in joint sale of produced polymers in the Russian and foreign markets.

3.2 Participation in the interstate power interconnections in NEA

(1) A general view on the problem

The interstate integration and cooperation between Russia and NEA countries in the field of electric power industry are the components of the Eastern energy strategy of the Russian Federation.

This cooperation suggests the development of interstate power interconnections between eastern regions of Russia and NEA countries.

This concerns a potential interconnection of power systems of Russia's East (East Siberia and the Far East), Mongolia, Chinese People's Republic, North Korea, South Korea and Japan for parallel operation.

The area served by such an interstate power super grid can contain various types of power plants (thermal, hydro, nuclear, wind, etc.) with a total installed capacity above 500 million kW.

(2) Interstate power interconnection in NEA: prospects for Russia

The electric power potential of East Siberia and the Far East is great. These regions have above 20% of the installed capacity of all power plants in Russia. The electric power export potential of eastern regions that can be realized in the near future according to different estimations makes up 10-15 million kW, including 6-7 million kW in East Siberia and 4-8 million kW in the Far East.

The calculations show that the reconstruction of existing power plants, completion of started construction of power plants, and commissioning of those planned for construction in the near future in East Siberia alone will provide a large amount of surplus power (according to various estimations from 30 to 40 TWh).

Such an amount of electric power can be transmitted by high voltage transmission lines in southeastern direction (Mongolia, China), and to Russia's Far East. Construction of the high voltage transmission line "East Siberia – Far East" will make it possible to integrate the interconnected power systems of Siberia and the Far East for parallel operation, which will enhance the reliability of power supply to consumers in these regions and create necessary preconditions for the construction of an Eastern "wing" of global electric power system [15, 16].

Currently, Russian companies are studying the possibility of constructing export-oriented power plants and high voltage transmission lines in the territory of Russia jointly with foreign companies to export a large amount of power (60-70 billion kWh) to the Republic of Korea, and 25-30 billion kWh – to Japan. Russia and China can actively and fruitfully cooperate with Mongolia to build the interstate electric power system "Russia – Mongolia – China" [17].

Russian load serving entities, research and design institutes are actively participating in the development of the Asian Super Grid concept [18].

Russia can actively and effectively cooperate with Mongolia in the construction of Gobitec and use of its solar and wind potential to build Asian Super Grid [19].

3.3 Innovation and technology cooperation in the field of coal Cooperation in advanced processing of brown coal

A promising direction in the innovation and technology cooperation between Russia and NEA countries in the field of coal industry is the establishment of joint ventures for complex, advanced processing of Russian brown coals and the creation of a new industry in the east of Russia in the form of coal chemical production clusters.

Brown coal reserves in East Siberia and the Far East make up more than 85 billion t [20]. The major brown coal deposits include:

- The deposits of Kansk-Achinsk basin in Krasnoyarsk Territory;
- The Svobodninskoye and Sergeyevskoye deposits in Amur Region;
- The Kangalasskoye deposit – in the Republic of Sakha (Yakutia);
- The Ushumskoye – in the Jewish Autonomous Region;
- The Mukhinskoye – in Khabarovsk Territory;
- The Solntsevskoye – in Sakhalin Region.

The cost of coal production is relatively low, therefore, these deposits can be used to set up joint ventures with foreign companies, that will be involved in the advanced processing of brown coal.

The products of advanced coal processing (such as motor fuel) can be supplied to consumers in Russia and other countries.

3.4 Cooperation in reliable power supply to off-grid and hard-to-access consumers in the eastern regions of Russia through the construction of renewable energy sources

According to the draft Russia's energy strategy 2035 [21], it is planned to place into service a great amount of renewable energy sources throughout this period. The projected increase in the share of renewable energy sources in the total electricity generation makes up from current 0.2 % to 3% in 2035. This is particularly important for the economy of the eastern regions.

The use of renewable energy sources (mini-HPP, Geo-TPP, solar and wind power plants) is a strategic priority in the energy development of northern and hard-to access territories of Russia's East.

We estimate the total renewable capacity to be placed into service in the off-grid areas and the areas with unstable power supply in the east of Russia at 600-870 MW until 2035 according to the base and moderate-optimistic scenarios, respectively [2]. The main share of renewable capacity to be commissioned in East Siberia falls on solar power plants (287-318 MW). The main share of capacities to be put into operation in the Far East is represented by wind turbines (84-155 MW) (Fig.1).

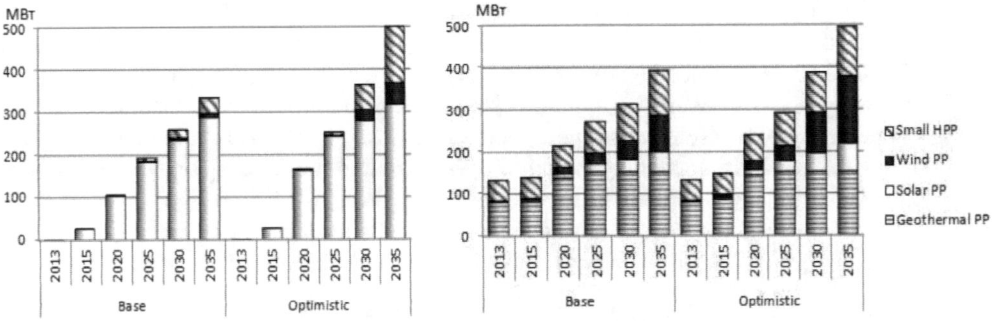

a) East Siberia b)Far East

Fig.1. Dynamics of installed renewable capacities

The total installed capacity of renewable energy sources in the eastern regions will rise by 5.5-7.5 times by 2035 and will make up 727-1000 MW including solar power plants – 335-385 MW and wind power plants -100-210 MW.

The Kamchatka Territory, Kuril islands of the Sakhalin Region, arctic coast of Krasnoyarsk Territory and the Republic of Sakha (Yakutia), Eastern part of the Magadan Region, Khabarovsk and Primorie Territories, and Northeast of Chukot Autonomous Area are the most promising for the development of wind power industry in Russia's East.

The priority areas for the development of solar power are the territories of the Republics of Buryatia, Tyva, Khakassia, Sakha (Yakutia); Transbaikal Territory; and southern areas of Irkutsk and Amur Regions.

The beneficial cooperation with foreign companies can be expected in the field of construction of renewable energy sources and establishment of joint ventures in the eastern regions of Russia.

The projects for construction of wind power plants in the settlement of Ust-Kamchatsk of Kamchatka Territory and settlement of Tiksi of the Republic of Sakha (Yakutia) with the Japanese companies Komaihaltec Inc. and MITSUI & Co. Ltd can serve as an example of such cooperation.

4 Conclusions

1. Implementation of Russia's Eastern energy strategy is a challenging task. This challenge lies in the complexity of the task, which is carried out on a vast territory, and involves a great number of participants from different countries. Moreover, the policy measures (especially interstate energy projects) that suggest tight international energy cooperation at the federal (state), regional and interregional levels are very capital intensive.

2. There is an urgent necessity to develop a scientifically grounded strategy (road map) of energy cooperation between Russia and NEA countries, which should show the order of energy resources development, sequence and time of their supply to domestic consumers, amounts of export and import deliveries. The strategy should assess socioeconomic and other consequences from specific interstate projects not only for the individual companies, but also for the regions and an entire country.

3. To promote mutually beneficial forms of energy cooperation between Russia and NEA countries it is necessary to meet at least the following five conditions:

1) There should be political will and serious intentions of participants to implement a specific energy project mutually beneficial for each country.

2) The economic and energy policy of central and regional authorities, and business community of the countries should be coordinated to develop interstate energy projects.

3) The consequences (effects) of the large-scale interstate energy projects for countries, regions and energy companies should be comprehensively assessed, considering high uncertainty of future development, economic risks and global challenges.

4) The mutually acceptable mechanisms (organizational, economic, legal and other) should be developed to implement interstate energy projects.

5) The interstate projects should be developed and done by an international team (in all the stages: from feasibility study and design works to their implementation).

References

1. Russia's energy strategy 2020, (approved by the resolution of the Government of RF No. 1234-p of August 28, 2003). 118 p., Moscow (September 2003)
2. Russia's energy strategy 2030 (approved by the resolution of the Government of RF No.1715-p of November 13 2009). http://www.energystrategy.ru/projects/es-2030.htm
3. Program for creation of unified gas production, transportation and supply system in East Siberia and the Far East, considering potential gas export to the markets of China and other APR countries (approved by the order No. 340 of Ministry of Industry and Energy of RF of 03.09.2007).
4. Strategy for socioeconomic development of the Far East and Baikal region 2025 (approved by the resolution of the Government of RF No.2094-p of December 28,2009) – http://www.minregion.ru/activities/territorial_planning/strategy/federal_development/346/
5. Strategy of socioeconomic development of Siberia 2020 (approved by the resolution of the Government of RF No.1120-p of July 5 2010).
6. Eastern vector of Russia's energy strategy: current state, look into the future. Ed. by N.I.Voropai, B.G.Saneev. Novosibirsk: Academic Publishing House "Geo",368 p. (2011)
7. *Fuel and energy complex of the Sakhalin region: current state and development prospects* /Ed. by B.G.Saneev, V.N.Tikhonkikh. Moscow: Publishing House "Energia", 240 p. (2010)
8. *Fuel and energy complex of the Amur region: current state and development prospects* /Ed.by B.G. Saneev, V.E.Peskov. Moscow: Publishing House "Energia", 240 p. (2010)
9. Energy strategy of the Republic of Sakha (Yakutia) 2030 /Ed. by G.F.Alekseyev, V.I.Kondratieva, A.A.Struchkov, N.A.Petrov, B.G.Saneev, A.F.Safronov. Yakutsk-Irkutsk: Media-Holding "Yakutia", 328 p. (2010)
10. *Fuel and energy complex of the Irkutsk region. Current state, development prospects* / Ed. by B.G.Saneev, P.A.Voronin. Moscow: Publishing House "Energia", 230 p. (2013)
11. *Energy sector of the Baikal region: current state, prospects, development strategy* /Ed. by B.G. Saneev. Novosibirsk: Academic Publishing House "Geo", 176 p. (2015)
12. A.A.Makarov, N.I.Voropai, V.V.Saenko , Saneev B.G., et al. *Regional features of fuel and energy complex* // in the book "Russia's energy: look into the future". Materials for the energy strategy of Russia 2030 /Ed. by A.B.Yanovsky et al. M. Publishing House "Energiya", p.435-460. (2010)
13. B.Saneev, D.Sokolov *Russian energy development in Eastern Siberia and the Far East and relations with East Asian countries in the energy sector* / Russia and East Asia.

Informal and gradual integration / Ed. by Tsuneo Araha and Anna Vassilieva, p. 181-199. (2014)

14. Plan for the development of gas and petro chemistry in Russia until 2030 (stage II), approved by the order of the Ministry of Energy of Russia No.79, (March 1, 2012).
15. N.I.Voropai, V.V.Ershevich, Y.N. Rudenko *Development of international power interconnections - a way to the creation of global electric power system*. Irkutsk: SEI SB RAS, 29 p. (preprint) (1995)
16. L.S.Belyaev, S.V.Podkovalnikov, V.A.Saveliev, L.Y.Chudinova. *Effectiveness of interstate electric ties*. Novosibirsk: Nauka, 239 p. (2008)
17. N.I.Voropai, B.G.Saneev, S.Batkhuyag, Kh.Enkhjargal. *Energy cooperation between Mongolia and Russia: current state and strategic directions.* Khabarovsk: Spatial economics, No.3, p.108-122. (2013)
18. V N.I.oropai, S.V.Podkovalnikov, B.G.Saneev. *Interstate cooperation in Northeast Asia: state, potential projects, energy infrastructure.* Moscow: Energy policy, No.2, p.55-64. (2014)
19. *Gobitec and Asian Supergrid for Renewable Energies in Northeast Asia* / S. Mano, B. Ovgor, Z. Samadov, D. Sokolov, B. Saneev, etc. Brussels, 110 p. (2014)
20. Program for the coal industry development in Russia 2030 /Ministry of Energy of the Russian Federation, Moscow, 178 p. (2014)
21. Draft energy strategy of the Russian Federation 2035 (edition of 01.02.2017) /Ministry of Energy of Russia, 78 p.
22. B.G.Saneev, I.Yu.Ivanova, T.F.Tuguzova. *Development of renewable energy in Russia's East in the first half of the 21st century against the background of the nationwide trends*. Energy policy, Iss.3, p.66-73. (2016)

Ensuring energy security in ASEAN countries: Current trends and major challenges

Sergey Senderov[1*], *Sergey* Vorobev[1]

[1]Melentiev Energy Systems Institute, 130 Lermontov str., Irkutsk, Russia

Abstract. The paper discusses the issues of formation of future challenges to energy security of the ASEAN countries in the period up to 2035. The article gives examples of strategic threats to the energy security of Russia. The opportunities to meet future demand for primary energy for individual countries of ASEAN and the whole region are discussed.

1 Introduction

Energy security (ES) in terms of economy and population of a country is understood as the condition of being protected against a threat of shortage of economically accessible fuel and energy resources (FER) of acceptable quality and against a threat of stable fuel and energy supply disruption. For normal operation of the fuel and energy complex (FEC) this condition of being protected implies, in a long-term context, satisfaction of the country's demand for electricity, thermal energy, boiler and furnace fuel and motor fuel. During extraordinary situations, for example, heavy emergencies at FEC facilities, the condition of being protected supposes guaranteed satisfaction of the minimum necessary energy demand of the country's vitally important consumers including those from the social sphere.

For the countries with scarce fuel and energy resources (Japan, Korea, Thailand, etc.) an important component of ES is reliability and guarantee of FER supply, and for the countries that are more or less provided with the resources (the USA, Great Britain, China, Vietnam can also be referred to as these countries) - it is energy independence: capability to cope with loss/decrease in FER supplies using their own resources.

Analysis of the current status and trends of Russia's FEC development reveals a number of strategic threats to ES even for Russia – the country which is provided with energy resources quite well. The notion of "strategic threats to the national ES" should be understood as the threats whose realization may cause a persistent and long-term shortage of energy in the country, which can retard economic growth of the country, lead to stagnation or deterioration of its social and economic situation. These strategic threats relate to such issues as: the lack of investment in the energy sector; high energy intensity of the economy; high wear equipment in the energy sector; the dominant role of natural gas in the European part of the country, etc.

* Corresponding author: ssm@isem.irk.ru

Russia does not have any particular problems of energy independence; it has enough reserves of energy resources now and in the future. What is most important to ensure energy security in the countries or regions of the world where energy resources are scarce?

Let's look at the situation with the prospects for energy security and the development of green energy in the ASEAN countries.

2 The share of nuclear and renewable energy sources in primary energy consumption in ASEAN

At the time in 2009-2016 the share of renewable sources, including hydro energy, in primary energy consumption of the world changed from 7,8% up to 10,0% mainly due to the increase in the use of other renewable resources. The data demonstrate the following situation in the world (Table 1).

Table 1. The share of renewable energy sources in primary energy consumption in the world, [1, 2]

	Primary energy	Hydro	Other renewable (without bioenergy)	Share of hydro	Share of renew. in Total
	mtoe			%	
2009	11326	738	143	6,5	7,8
2010	11956	784	168	6,6	8,0
2011	12232	796	205	6,5	8,2
2012	12483	834	241	6,7	8,6
2013	12730	856	279	6,7	8,9
2015	13105	883	366	6,7	9,5
2016	13276	910	419	6,8	10,0

The same situation is observed in the world regions. For example for the entire North America this share increased only by 1.3% (from 13.4% to 14.7%), for Europe and Eurasia – by 1.4% (from 16.7% to 18.1%). In the regions of South and Central America, Middle East, Africa and APR this share did not change at all and made up 28.3%, 0.4%, 7.3% and 9%, respectively. The data [1] demonstrate the following situation in the largest ASEAN countries (Table 2).

Table 2. The share of renewable energy sources in primary energy consumption in ASEAN, [1, 2]

	Primary energy	Hydro	Other renewable	Share of hydro	Share of renew. in Total
	mtoe			%	
2009	485	14,8	128,9	3,1	29,6
2010	523	14,8	140,2	2,8	29,6
2011	549	18	139,2	3,3	28,6
2012	568	21,2	140,1	3,7	28,4
2013	594	21,3	143,7	3,6	27,8
2015	562	22,4	144,8	3,9	29,7
2016	589	24,1	149,1	4,1	29,4

Analysis of Table 2 shows that, compared with the global trends (also not very fast, 1% for 7 years) the process of change in the share of renewable energy in total energy consumption has a reverse trend. Over the past seven years, this share decreased by 0,2%. In general, the decrease can be explained by a gradual increase in the share of centralized power and correspondingly by reducing the share of biomass use by individual consumers in rural areas.

Now we can consider how this process has been in major ASEAN countries in the past five years, Table 3.

Table 3. The share of renewable energy sources in primary energy consumption in largest ASEAN countries, [1, 2]

	2009	2010	2011	2012	2013	2015	2016
Thailand							
Primary energy, mtoe	95,9	102,4	106,4	113,7	115,7	121,8	123,8
Hydro & oth.ren., mtoe	22,1	22,4	22,7	23,1	23,5	23,7	24,1
Share of renewable, %	***18,7***	***18,3***	***17,8***	***17,5***	***17,2***	***17,5***	***17,9***
Malaysia							
Primary energy, mtoe	73,6	72,4	79,8	83,2	89,2	93,8	99,5
Hydro & oth.ren., mtoe	1,9	1,7	2,2	2,4	2,9	3,8	4,5
Share of renewable, %	***2,5***	***2,3***	***2,7***	***2,8***	***3,2***	***4,1***	***4,5***
Singapore							
Primary energy, mtoe	64,3	69,0	71,7	72,0	74,1	81,0	84,1
Hydro & oth.ren., mtoe	0,1	0,1	0,1	0,1	0,2	0,2	0,2
Share of renewable, %	***0,2***	***0,1***	***0,1***	***0,1***	***0,3***	***0,2***	***0,2***
Indonesia							
Primary energy, mtoe	200,3	206,1	211,9	217,2	222,3	224,7	235,0
Hydro & oth.ren., mtoe	54,8	55,9	57,1	58,3	59,5	59,9	60,0
Share of renewable, %	***27,3***	***27,1***	***26,9***	***26,8***	***26,8***	***26,6***	***25,5***
Vietnam							
Primary energy, mtoe	39,3	44,3	50,3	52,5	54,8	63,7	64,8
Hydro & oth.ren., mtoe	6,8	6,2	9,3	12,0	12,9	12,9	13,8
Share of renewable, %	***17,3***	***13,9***	***18,5***	***22,8***	***23,5***	***20,2***	***21,3***
Philippines							
Primary energy, mtoe	33,5	36,7	39,8	42,8	46,2	51,7	57,1
Hydro & oth.ren., mtoe	15,4	16,0	16,8	17,5	18,2	18,7	20,1
Share of renewable, %	***46,0***	***43,6***	***42,2***	***40,9***	***39,4***	***36,1***	***35,2***

In Thailand the share of renewable energy in the past seven years was decreased by 0,8%. In Singapore this share has remained at the same level as that of 7 years ago. In Indonesia, it decreased by 1,8% in the Philippines decreased by 10,8%. Only in Malaysia and Vietnam, this share was increased. For Vietnam, this is due to an active construction and commissioning of hydropower plants.

Comparison of data from Table 3 makes it possible to assess changes in the dependence of the largest ASEAN countries on supplies of the main types of non-renewable primary energy: coal, heavy oil, natural gas in 2009-2016 (Table 4).

Table 4. Relative change in the values of primary energy consumption and in the average share of renewable resources in this consumption in the largest ASEAN countries

Country	Change 2016 over 2009, %	
	Primary energy consumption	Share of renewable energy resources
Thailand	29,1	-0,8
Malaysia	35,3	2,6
Singapore	30,7	0,0
Indonesia	17,3	-1,8
Vietnam	64,8	4,0
Philippines	70,4	-10,8
ASEAN as a whole	***21,4***	***1,4***

Analysis of data from Table 4 shows a dynamic increase in consumption of non-renewable primary energy in all ASEAN regions in comparison with reducing in the share of renewable energy in the total primary energy consumption.

3 Dependence of the largest ASEAN countries on imported primary energy

Now let us focus on the non-renewable primary energy resources consumed to produce different types of energy. Table 5-7 show how the demand of primary energy can be covered in the countries – major energy consumers in ASEAN.

3.1 Dependence on imported natural gas

Table 5 shows information about main existing trends concerning of the dependence of the largest ASEAN countries on imported natural gas.

Data in Table 5 show that Singapore fully depends on external supplies of natural gas, 20% is depending of Thailand. Given the production of natural gas in Brunei and Myanmar about 26 million toe in total, it is seen that the region is balanced on production and consumption of natural gas. Gas flows in the region in 2016 were supplied through the main gas pipeline from Indonesia to Singapore, from Malaysia and Myanmar to Thailand and from Brunei to Singapore. Shipments of liquefied natural gas in 2016 were carried out mainly from Qatar, Africa and Brunei.

Table 5. Dependence of the largest ASEAN countries on imported natural gas, [1, 2]

Country	Production			Consumption			Dependence on import		
	2010	2013	2016	2010	2013	2016	2010	2013	2016
	Million toe						%		
Thailand	32,7	37,6	34,7	40,6	47,0	43,5	19	20	20
Malaysia	58,8	62,1	66,5	31,6	30,6	38,7	-	-	-
Singapore	-	-	-	7,6	9,5	11,3	100	100	100
Indonesia	73,8	63,4	62,7	36,3	34,6	33,9	-	-	-
Vietnam	8,5	8,8	9,6	8,5	8,8	9,6	-	-	-
Philippines	3,2	3,0	3,4	3,2	3,0	3,4	-	-	-

3.2 Dependence on imported coal

Table 6 shows information about main existing trends concerning of the dependence of the largest ASEAN countries on imported coal.

Table 6. Dependence of the largest ASEAN countries on imported coal, [1, 2]

Country	Production			Consumption			Dependence on import		
	2010	2013	2016	2010	2013	2016	2010	2013	2016
	Million toe						%		
Thailand	5,1	5,1	4,3	15,8	16,0	17,7	68	68	75
Malaysia	1,9	2,0	2,0	14,8	17,0	19,9	87	88	90
Singapore	-	-	-	-	-	0,4	-	-	100
Indonesia	169,2	258,9	255,7	41,2	54,4	62,7	-	-	-
Vietnam	24,6	23,1	22,0	13,9	15,9	21,3	-	-	-
Philippines	5,2	5,2	5,2	7,7	10,5	13,5	32	50	61

Data in Table 6 show that Malaysia, Singapore, Thailand and Philippines are significantly depend on external supplies of coal. At the same time, a significant excess of coal production over its consumption in Indonesia completely covers the demand for external coal supplies from these countries.

Comparison of possibilities of primary energy production and scale of their consumption by country (Table 5-6) shows the countries that depend on the external energy supplies to the greatest extent. This is Singapore. The rest of the largest ASEAN countries can be subdivided into the following groups:

- the countries that satisfy their demand for gas and coal by local resources and export them (Indonesia, Vietnam);
- the countries that partly import some resources (Thailand imports some amount of gas and coal, Malaysia - some amount of coal) but have equivalent production reserves of natural gas. Philippines partly import coal, but completely consume only its own natural gas.

On the whole, if not to consider oil fuels, the ASEAN region can be considered as energy resources self-sufficient region.

If you try to piece together the possible production of natural gas and coal, and their consumption, you can get the total picture of the dependence of the largest countries in ASEAN from these important primary energy resources.

Table 7. Dependence of the largest ASEAN countries on imported natural gas and coal

Country	Production gas and coal			Consumption gas and coal			Dependence on import		
	2010	2013	2016	2010	2013	2016	2010	2013	2016
	Million toe						%		
Thailand	37,8	42,7	39,0	56,4	63,0	61,2	33	33	36
Malaysia	60,7	64,1	68,5	46,4	47,6	58,6	-	-	-
Singapore	-	-	-	7,6	9,5	11,7	100	100	100
Indonesia	243,0	322,3	318,4	77,5	89,0	96,6	-	-	-
Vietnam	33,1	31,9	31,6	22,4	24,7	30,9	-	-	-
Philippines	8,4	8,2	8,6	10,9	13,5	16,9	23	39	49

Data of Table 7 show that Indonesia, Malaysia and Vietnam can be characterized as independent country in terms of primary energy consumption. Thailand, the Philippines and Singapore are in varying degrees dependent on external supplies of these kinds of energy and can't cover its energy needs from their own sources. Such situation in the Philippines is rapidly getting worse.

3.3 Dependence on imported oil

According to the information of the World Energy Agency [2], by 2030 consumption of primary energy in the world may increase up to 17 billion toe and in the Asian region – up to 7 billion toe. Despite this fact the anti-nuclear sentiments related to the problems of nuclear energy security, radioactive waste storage and burial are still very strong. The situation was to a great extent exacerbated after the emergency at the Fukushima NPP. Thus, the main load in the growing consumption of primary energy again falls on traditionally used gas, oil and coal.

Almost all oil in ASEAN countries after refining is used for transport purposes, therefore, we will consider the situation with the dependence of the largest ASEAN countries consuming energy on external oil supplies, Table 8.

Table 8. Dependence of the largest ASEAN countries on imported oil, [1, 2]

Country	Production			Consumption			Dependence on import		
	2010	2013	2016	2010	2013	2016	2010	2013	2016
	Million toe						%		
Thailand	14,8	16,6	17,6	44,3	50,4	59,0	67	67	70
Malaysia	32,0	29,6	32,7	29,3	31,2	36,3	-	5	10
Singapore	-	-	-	62,8	65,9	72,2	100	100	100
Indonesia	48,6	42,7	43,0	66,4	73,8	72,6	27	42	41
Vietnam	15,3	17,0	16,0	15,6	17,4	20,1	2	2	20
Philippines	1,3	1,2	1,2	13,2	13,7	19,9	90	91	94

Data from Table 8 are indicative of the reliance of the largest ASEAN countries on imported oil (Singapore and Philippines are almost totally dependent). Fig.1 shows the tendencies towards increase in the dependence of some countries and region as a whole. In previously independent of external supplies of oil Malaysia for the past 4 years the dependence on external supplies of oil increased up to 5%. Also up to 42% (in one and a half) dependence on external supplies of oil has increased over the last 4 years in Indonesia.

In general, the ASEAN region is deficient in the oil and significantly depends on its external supply. Structure of net oil imports in countries that depends from external oil supply is roughly similar to structure that shown in Fig. 2 for Singapore in 2016.

 https://doi.org/10.1051/e3sconf/20182702002

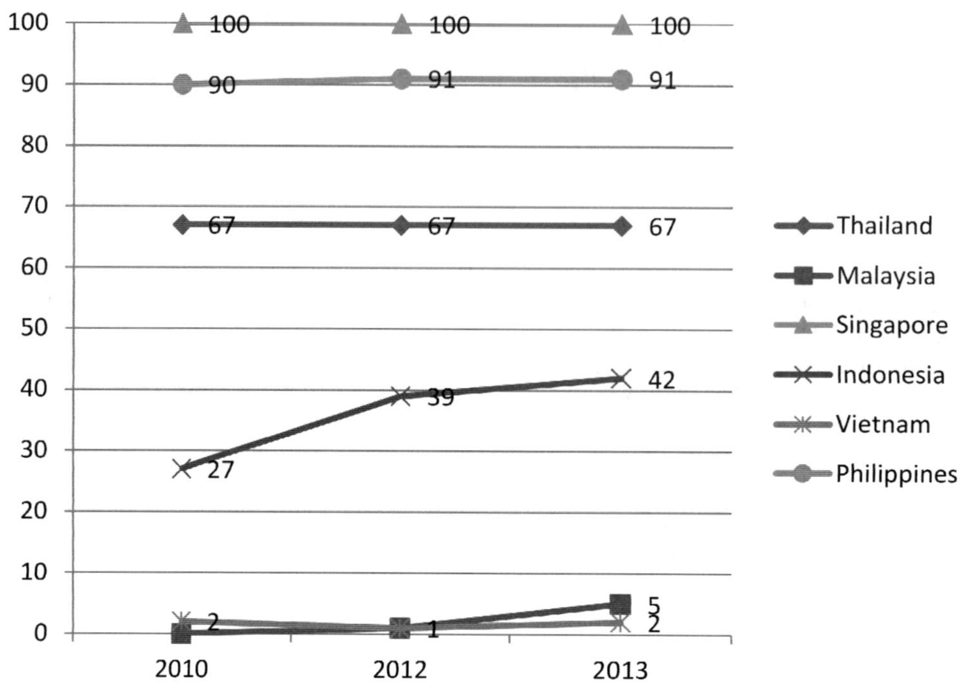

Fig.1. Change in dependence of the largest ASEAN countries on external oil supplies

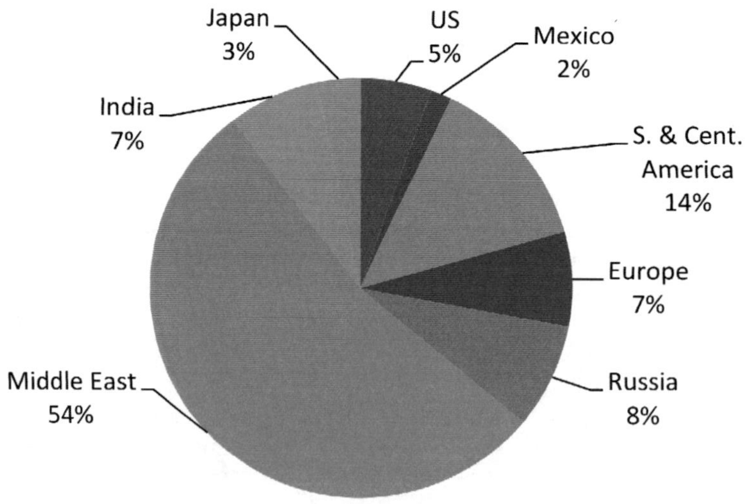

Fig. 2. Structure of Singapore's net oil import in 2016, [1]

Data of Fig. 2 show that the structure of the external oil supply to Singapore is: more than half of such deliveries made from the Middle East and the rest are more or less evenly divided between the 7 countries - oil exporters.

As seen from table 8, ASEAN countries actively increase their dependence on imported oil. This can greatly reduce the energy security of the region in the run up to 2035, when it may cause serious decline of oil production in the existing fields on the background of insufficient pace of development of new oil fields. IEA estimates that by 2035 the overall decline in oil production could reach about 30% of the existing oil production volumes. At the same time, according to some estimates ASEAN could become fourth largest oil importer after China, India and the countries of the EU. Region's dependence on oil by that time could double and increase to 75%, and net imports may increase from 2 million barrels per day to 5 million barrels per day. Expenses of the region on oil imports could triple and reach about 250 billion US dollars in constant prices. More than 50% of this amount will likely have on Indonesia and Thailand [2].

IEA estimated proved oil reserves at the beginning of 2014 about 1.8 billion tons. At the current level of oil production, these reserves can be used for approximately 14 years.

3.4 The main energy characteristics of the ASEAN countries

So, we can formulate the main energy characteristics of the ASEAN countries. Based on this analysis, it is possible to comment on the main energy characteristics for each of the countries.

Indonesia is a largest energy consumer in ASEAN with significant potential for further growth in energy consumption. It is a largest exporter of coal, as well as exporter of LNG [3].

Thailand is the second largest energy consumer in ASEAN. It is largely depends on energy resources imports [4].

Malaysia is the third largest energy consumer in ASEAN. It is largest exporter of LNG, until recently exported oil, but since 2012 it has become an importer and dynamically increases the volume of oil imports [5].

Singapore is a key node of trade and oil refining in ASEAN. It is entirely depends on external supplies of primary energy resources (oil, natural gas) [6].

Vietnam has significant resources of renewable energy, fully satisfies their needs by its own coal, gas and oil almost completely. Vietnam actively develops nuclear power and oil refining [7].

Philippines consume only its own natural gas, are completely depends on imports of oil and partly on imported coal. In this country rapidly grows electricity demand [8].

Myanmar has rich resources of gas and significant hydropower potential. Great potential for growth in energy consumption in the country is necessary for economic growth and improving living standards [9].

Laos has significant plans to develop hydropower on Mekong River and characterized by the growth of electricity exports. But construction of hydroelectric power stations on the Mekong River, which flows around the peninsula of Indochina, will reduce the amount of fish that is a staple in the diet of the inhabitants of Laos, Thailand, Cambodia and Vietnam. This will force more to raise cattle, increased costs of water, which is very difficult for these countries and for their nature, according to the World Wildlife Fund [10, 11].

Cambodia is characterized by a low level of electrification. There is potential for the detection of oil and gas reserves in Cambodia [12].

Brunei is one of the first in Asia in terms of life. Base of its economics are oil and natural gas. Exports of this resources generates for about 60% of Brunei GDP [13].

4 Expected demand for primary energy in ASEAN

According to the New Policies Scenario, primary energy consumption in ASEAN by 2035 could increase by almost 80% compared with the current year and will be about 1 billion toe, Table 9. The growth rate slows progressively from an average of 3% per year to 2020 to 2.3% per year from 2020 to 2035, largely reflecting a gradual decline in economic and population growth rates, plus results from policy efforts to move towards more efficient patterns of energy use. The region remains heavily dependent on fossil fuels, which collectively represent 80% of primary energy demand in 2035.

Table 9. Primary energy demand in the ASEAN countries for the period until 2035, Mtoe. Based on [2]

	2013	**2016**	**2020**	**2025**	**2035**
Indonesia	222	232	255	285	360
Malaysia	83	101	95	110	130
Philippines	46	55	60	70	95
Thailand	137	147	150	170	210
Rest of ASEAN	152	155	160	180	220
Total ASEAN	**640**	**690**	**718**	**804**	**1 004**

Table 9 and Fig. 3 show that in general the consumption of primary energy in the region from 2013 until 2035 could grow by almost 60%. It is important to note that the volume of use of nuclear energy, hydropower, bioenergy and other renewable energy resources is increasing very moderately. In 2035 in total, these resources will not take more than 20% of total primary energy consumption in the region. Accordingly, the consumption of oil, gas and coal, according to the construction of scenarios, grows rapidly and can reach an 80% (today - 72%) in the total primary energy consumption.

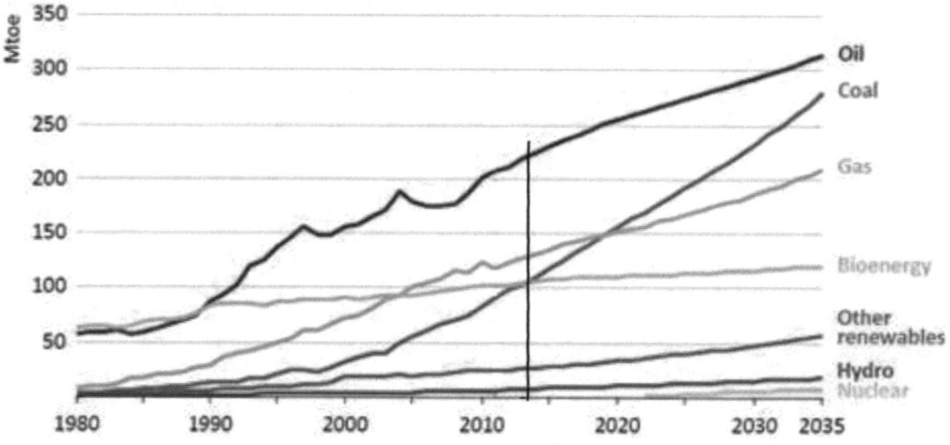

Fig 3. ASEAN primary energy demand by source [2]

About nonrenewable energy resources in this case we can say the following. Oil consumption in the region during the analyzed period (by 2035) may increase by 1.5 times. As before, the oil will hold the largest share in the primary energy consumption in the region.

Coal demand triples, growing at 4.8% per year on average over the projection period. It overtakes natural gas from 2020 to become the second-largest component of Southeast Asia's energy mix, coal's share reaching 28% in 2035. This rapid increase in the share of coal in primary energy consumption, of course, should be explained by the availability of sufficient stocks of coal production (especially in Indonesia) and by logical desire to replace part of consumed oil and natural gas on that coal especially for new capacities.

Natural gas consumption in the region could rise to 220 bcm in 2035 increased compared to the present day by almost 70%. The share of its total consumption of primary energy will remain until the same level (about 20%). This is explained by the rather high price of local natural gas and LNG imported into the region. Perhaps the situation will change with the increasing demands on the limitation of emissions of carbon dioxide, which will slow down the consumption of coal in favor of gas.

The share of renewables (including traditional biomass) in Southeast Asia's primary energy demand falls from 28% in 2013 to 20% in 2035. This results from a large shift away from the use of traditional biomass (mostly fuel wood, charcoal, animal dung and agricultural residues used for household cooking and water heating) in favour of modern fuels, which is driven by rising living standards and ongoing urban migration. The share of traditional biomass in the region's primary energy demand drops from 11% in 2013 to 6% in 2035. Countering this trend is growth in demand for modern renewables – including geothermal, hydro, wind, solar and modern biomass.

Nuclear power does not currently feature in the energy mix of any Southeast Asian country. There is concern about the future of energy security, and accordingly there are plans to enter the nuclear power plant in Vietnam, Thailand, but it will be after 2020.

Now we can consider the possible trends in the energy development of some ASEAN countries from the positions of energy security.

Indonesia. In the New Policies Scenario, Indonesia's total primary energy demand growth averages 2.5% per year over 2013-2035, rising from 222 Mtoe to nearly 360 Mtoe, Table 10. Over the period, its population expands from 242 million to 302 million and its economy grows by around 220%.

Table 10. Primary energy demand in Indonesia by fuel (Mtoe). Based on [2]

	2013	**2016**	**2020**	**2025**	**2035**
Coal	54	63	60	78	115
Oil	74	73	84	87	95
Gas	35	34	46	53	71
Hydro	3,5	3,3	3,5	3,6	3,7
Bioenergy*	40	40	39	37	38
Other renewables	16	19	21	25	37
Total	**222**	**232**	**255**	**285**	**360**

* Includes traditional and modern biomass uses.

When analyzing the prospects for energy development in Indonesia (tab. 10), and also taking into account the already existing situation (Table. 5, 6, 8), you can see the following trends. Coal consumption by 2035 compared to 2013 should grow by 2.1 times, it is justified in terms of energy security because of the substantial reserves of the country for its production. Gas consumption during this period should also grow to 2 times. Now Indonesia has its own natural gas, but for the last three years, the volume of its production decreased by 14%. If such dynamics continues, by 2025 Indonesia will become a country - importer of natural gas. As for oil, in the last three years its production in the country has decreased by 12%, while consumption grew by 11%. Dependence on imported oil in Indonesia has increased during that time from 27% to 42%. Till 2035 oil consumption in Indonesia will increase by 28%. At the same time, the share of renewable energy in the structure of primary energy consumption can be reduced up to 20% in 2035 against 25% now. Emerging trends will lead to the need for import of natural gas, to an even greater dependence on oil imports and in general on energy imports, to reducing the possibility of future energy security providing.

Thailand. In the New Policies Scenario, Thai primary energy demand grows at 2.3% on average per year between 2013 and 2035, for an overall rise of 75% (Table 11). Key drivers include ongoing urbanization and a near tripling in the size of its economy. Per-capita energy use continues to increase, reaching 3 toe in 2035.

Table 11. Primary energy demand in Thailand by fuel (Mtoe). Based on [2]

	2013	2016	2020	2025	2035
Coal	16	18	28	34	47
Oil	50	59	61	63	74
Gas	47	44	48	50	51
Nuclear	-	-	-	-	2
Hydro	1,3	0,8	1,3	1,4	1,5
Bioenergy	22	23	26	27	30
Other renewables	0,2	2,8	0,2	0,5	0,9
Total	**137**	**147**	**160**	**180**	**210**

Prospects of energy consumption development of Thailand (tab. 11), and already existing trends (Table. 5, 6, 8) show the following situation. Coal consumption by 2035 compared to 2013 should grow by almost 3 times, but dependence on imported coal in Thailand is already at 70%. Gas consumption for this period may grow by 9%. Gas production in the country is growing (for the last 3 years growth of its production was 15%), but even more rapidly growing demand for natural gas (17% for the same 3 years). Current dependence on gas imports in Thailand is about 20% and it seems that this level of dependence will persist and continue. As for oil, in the last three years, the country's dependence on its imports remained at 67%. By 2035, the country's oil demand may rise more in the half. At the same time, the share of renewable energy in the total primary energy consumption could fall from 17% to 15%. Positive expectations are connected to input nuclear electricity generation capacity by 2030 (if it happens). In general, emerging trends gradually lead the country to a further increase dependence on energy imports, reducing the possibility of providing perspective energy security.

Philippines. In the New Policies Scenario, primary energy demand in the Philippines doubles by around 2030 and then rises further to 92 Mtoe in 2035, at a growth rate of 3.5% per year on average (Table 12).

Table 12. Primary energy demand in Philippines by fuel (Mtoe). Based on [2]

	2013	2016	2020	2025	2035
Coal	11	13	16	20	28
Oil	14	20	21	25	28
Gas	3	3	5	6	9
Hydro	2,2	2,1	2,4	2,6	3
Bioenergy	7	7	8	8	8
Other renewables	9	10	12	14	18
Total	**46**	**55**	**60**	**75**	**95**

Prospects of energy consumption development of the Philippines (tab. 12), and also already existing trends (Table. 5, 6, 8) shows the following features. Coal consumption by 2035 compared to 2013 should grow by 2.5 times. The volume of coal production in the country is insufficient, but coal consumption of the past three years has increased by 36%. Dependence on imported coal in the country is already at 50%. Gas consumption for this period can grow three times. At the same time, it is unclear what volume of gas production can the Philippines reach during the analyzed period. While the country is consumed only own gas that is produced there. As for oil, its production is relatively small but in the last three years its consumption was steadily increased. By the end of 2013, the dependence on imported oil in the country was 91%. By 2035, the country's oil demand may rise twice. The share of renewable energy in the total primary energy consumption, are likely to remain at the same level (29-30%). In general, emerging trends gradually lead the country to a further increase dependence on energy imports, to reducing of the opportunities for providing of energy security.

Malaysia. In the New Policies Scenario, Malaysia's primary energy demand may increase of 57% in 2013-2035 with annual average growth of 2.3% (Table 13). Growth in demand slows over time as growth in population and GDP moderate. Malaysia's per-capita energy consumption is currently relatively high for the region, at 61% of the OECD average.

Table 13. Primary energy demand in Malaysia by fuel (Mtoe). Based on [2]

	2013	2016	2020	2025	2035
Coal	17	20	25	30	35
Oil	31	36	37	38	40
Gas	31	39	40	42	45
Hydro	2,1	4,2	4,3	4,7	4,9
Bioenergy	2,2	2,3	2,8	3,2	4,4
Other renewables	0,0	0,3	0,9	0,9	1,0
Total	**83**	**101**	**110**	**118**	**130**

When analyzing the prospects for energy development of Malaysia (tab. 13), and the current trends (Table. 5, 6, 8) we can see the following features. Coal consumption by 2035 compared to 2013 should grow by 2.3 times. The volume of coal production in the country is insufficient, but the consumption of the past three years has increased by 15%. Dependence on imported coal in the country is already at 90%. Gas consumption for this period could grow by 30%, which could be covered by own sources of natural gas. As for oil, over the past three years with declining oil production and increasing consumption Malaysia moved into oil importer status. By the end of 2013, the dependence on imported oil in the country amounted to 5%. By 2035, the country's oil demand may rise 1.3 times. The share of renewable energy in the total primary energy consumption, are likely to remain at the same level (5-6%). In general, emerging trends gradually lead the country to a further increase dependence on energy imports, to reducing of the opportunities to provide energy security of the country.

5 Expected production of primary energy in ASEAN

In the New Policies Scenario, the outlook for fuel production varies by fuel and by country (fuel-by-fuel discussion follows). Southeast Asia's oil production falls by nearly one-third, from 2.5 million barrels per day (mb/d) in 2012 to 1.7 mb/d in 2035, with continued decline across the region's big mature oil fields (Figure 3.2). Prospects for gas production are brighter because of a larger base of remaining resources and strengthening demand across the Asia-Pacific market. Gas production in Southeast Asia increases from 203 bcm in 2011 to about 260 bcm in 2035, though growth slows after 2020. The region's coal output continues to expand, led by Indonesia, to meet fast-increasing domestic demand and export growth (though this occurs mainly in the medium term). ASEAN coal production increases from 348 Mtce in 2011 to around 620 Mtce in 2035.

Now we can consider the possible prospects of fossil fuels production in the region (Fig. 4).

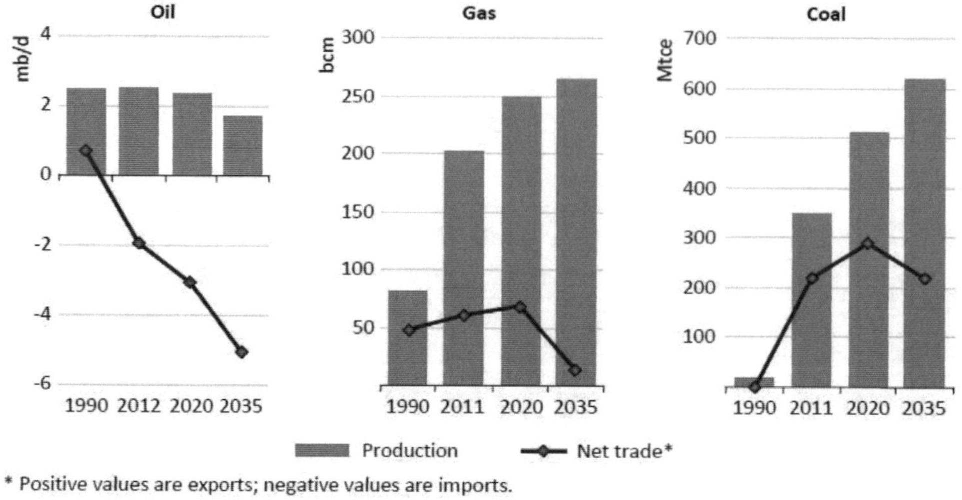

Fig. 4. ASEAN fossil fuel production and net trade [2].

The fig.4 shows that oil production will decline and by 2035 will fall to 1.7 mb/d (80 mtoe per year). This drop is due to a reduction in oil production on old oil fields. Taking into account the plans for oil consumption in 2035, by this time the region will has to import for about 5 mb/d of oil (240 mtoe per year).

Significant gas reserves in deposits of the region allow to hope for a significant increase in gas production by 2035 - about 260 bcm (235 mtoe) per year. Taking into account the plans for gas consumption in the region in 2035, the net gas exports could reach around 15 bcm (13-14 mtoe) per year.

According to the Scenario coal mining will actively develop in the ASEAN (Indonesia first of all). By 2035, the whole region should produce some 620 Mtce of coal (380 mtoe) per year. Taking into account the plans for coal consumption in the region in 2035, the net coal exports can be about 210 mtce (140 mtoe) per year.

Thus, the analysis of long-term consumption of primary energy and analysis of long-term production of this energy (Scenarios of new policies) let us make the following conclusions.

6 Conclusions

Despite attempts to increase the share of renewable energy in primary energy consumption structure, this share is reduced and by 2035 could reach only 20%.

ASEAN by 2035 will be 75% dependent on oil imports, but his own needs will be met by natural gas and coal. Turning to individual countries, we can say that all countries, except Brunei, will be significantly dependent on external oil supplies. All countries except Malaysia, Indonesia, Vietnam and Brunei will depend on external supplies of natural gas. All countries except Indonesia and Singapore, will be dependent on external supplies of coal. There are significant risks associated with the containment of nuclear power development (anti-nuclear sentiment in the world).

These challenges to the energy security of the ASEAN countries can be reduced only by a radical increase in the share of renewable energy in the structure of primary energy consumption, particularly in countries heavily dependent on energy imports.

In regard to oil, the strong dependence of the region on oil can be reduced only through the active increasing of the usage of biofuels, electric or hybrid vehicles.

References

1. BP Statistical Review of World Energy, (June 2017). Available: http://www.bp.com [Accessed 10 September 2017]
2. Southeast Asia Energy Outlook. World Energy Outlook Special Report, (2013). Available: http:// www.iea.org [Accessed 18 September 2017]
3. Energy Policies Beyond IEA Countries – Indonesia, (2015). Available: http://www.iea.org [Accessed 10 September 2017]
4. World Energy Outlook Special Report on Southeast Asia, (2015). Available: http://www.iea.org [Accessed 12 September 2017]
5. Malaysia Energy Statistics. Available: http://www.iea.org [Accessed 5 May 2017]
6. Singapore Energy Statistics. Available: http://www.iea.org [Accessed 10 May 2017]
7. Vietnam Energy Statistics. Available: http://www-pub.iaea.org [Accessed 10 September 2017]
8. Philippines Energy Statistics. Available: http://www.iea.org [Accessed 12 May 2017]
9. Myanmar Energy Statistics. Available: http://www.iea.org [Accessed 5 May 2017]
10. Future energy demand in Laos, (2012). Available: http://www.utu.fi [Accessed 18 August 2017]
11. The construction of a hydropower plant on the Mekong River in Indochina will deprive local fish residents, (2012). Available: http://ria.ru [Accessed 5 May 2017]
12. Cambodia Energy Sector Strategy. Available: http://www.un.org [Accessed 5 May 2017]
13. ASEM Ministerial Conference on Energy Security – Brunei. Available: https://eeas.europa.eu [Accessed 12 August 2017]

A comprehensive assessment of comparative effectiveness of projects for power export from East Siberia to China: A methodological approach and results of its application

Anatoly Lagerev[1,*], *Valentina* Khanaeva[1], and *Konstantin* Smirnov[1]

[1]Melentiev Energy Systems Institute, Irkutsk, 130 Lermontov St., Irkutsk, 664033, Russia

Abstract. The paper is concerned with a methodological approach to the assessment of comparative effectiveness of projects for the construction of export-oriented power plants and transmission lines under uncertainty of the power industry development in the region. The recommendations are given to select the most preferable project for the construction of an export-oriented power plant and transmission line for power export from East Siberia to China.

1 Introduction

Interstate integration and cooperation between Russia and NEA countries in the field of power industry is an indispensable part of the eastern energy policy of our country [1]. This cooperation suggests the development of interstate electric ties between the eastern regions of Russia and Chinese People's Republic, the Republic of Korea, Japan, Mongolia, and the Democratic People's Republic of Korea.

The transboundary transmission lines to be constructed for these purposes can be used both to export power from Russia, and to integrate power systems of the indicated countries for joint operation and formation of common electricity and capacity markets [2, 3].

In the future, the development of these ties is first of all planned through the projects of large-scale power export from Russia's eastern regions to the People's Republic of China [1,4,5].

An analysis of the export projects proposed in the eastern regions shows that their implementation (from design to completion of construction) is rather long (15-20 years), and is related to a considerable uncertainty of future conditions for the development of the power industry in the regions where these projects can be implemented (fuel prices, level of electricity consumption, etc.). In this situation, making decisions is subject to risk. Therefore, minimization of risks and study of various factors that affect the stability of decisions to be made under the considered conditions are essential for making decisions on export projects [6-11].

* Corresponding author: lagerev@isem.irk.ru

2 Description of a methodological approach

A methodological approach proposed in the paper to comprehensively assess comparative effectiveness of projects for the construction of export-oriented power plants and power transmission lines under uncertainty of the electric power system expansion in the region includes four stages (Fig.1):

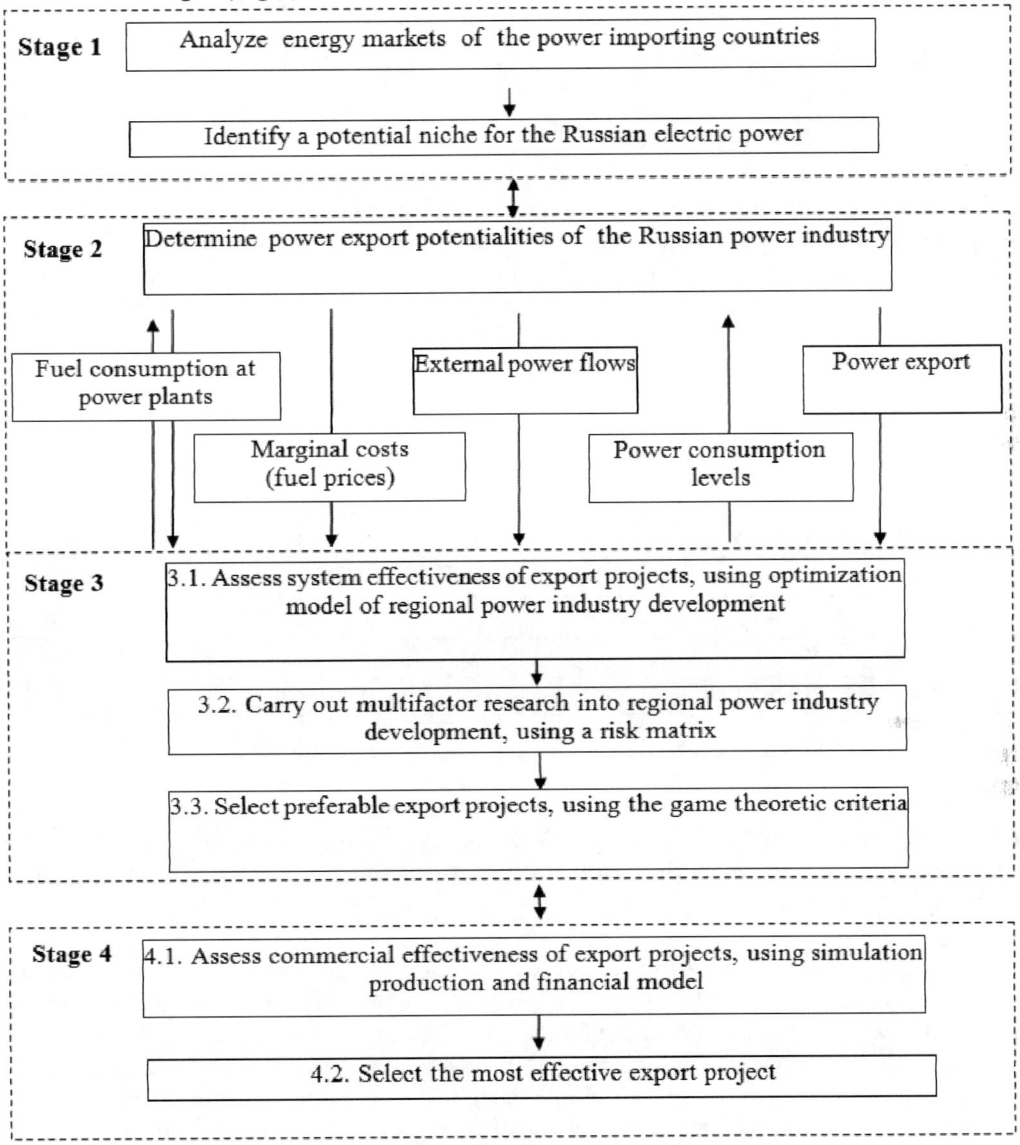

Fig.1. A block diagram of the methodological approach to comprehensively assess comparative effectiveness of power export projects under uncertainty of regional power industry development

1. Analysis of energy markets of the countries importing power, and determination of a potential niche for the Russian electric power.

2. Investigation of power export potentialities in terms of the national energy sector: fuel consumption at power plants of the region, fuel prices, external power flows between regions, volumes of power export from the region.
3. Assessment of the system effectiveness of export projects. To this end we:

- construct an optimization model of power system expansion in the region, and include alternately each considered project for the construction of export-oriented power plants and transmission lines in the region in this model.

- conduct a multifactor research into the regional power industry development, by using a payoff matrix and a risk matrix, which includes:

a) simulation of external conditions C^n of power system expansion (levels of power consumption, fuel prices, etc.);

b) determination of an optimal mix of energy facilities required for the expansion of regional power system for each of the considered export project, which is necessary to adjust the power system to the external conditions and planned power export deliveries;

c) construction of a payoff matrix $\| F_r^n \|$ (Table 1). In a general form, the coefficients of the payoff matrix F_r^n are defined as the total discounted costs of the regional power system expansion under the considered external conditions C^n of its expansion, and projects for construction of export-oriented power plants and transmission lines X_r. The obtained payoff matrix makes it possible to switch to the matrix of economic risks.

Table 1. A payoff matrix of regional power system expansion costs

Variants of power system expansion. Projects	**Conditions for the power system expansion**					
	C^1	C^2	C^3	C^n	.....	C^N
X_1	F_1^1	F_1^2	F_1^3	F_1^n	.....	F_1^N
X_2	F_2^1	F_2^2	F_2^3	F_2^n	.....	F_2^N
X_r	F_r^1	F_r^2	F_r^3	F_r^n	.....	F_r^N
.....	.....	.....	.....	.....	.....	.....
X_R	F_R^1	F_R^2	F_R^3	F_R^n	.....	F_R^N

d) construction of a risk matrix $\| \lambda_r^n \|$, whose coefficients are determined by subtracting the index of discounted costs F_{min}^n for the variant which under the conditions C^n has the minimum value of costs from the index of discounted costs F_r^n of the considered variant r of the payoff matrix, i.e.

$$\lambda_r^n = F_r^n - F_{min}^n$$

Coefficient of the matrix of economic risks λ_r^n is a characteristic of extra costs of the considered variant of regional power system expansion under certain conditions compared to the variant with the minimum costs under these conditions.

In a general case, the matrix of economic risks represents a generalized characteristic of possible economic consequences of the ignorance of real conditions for the regional power system expansion, and is a basis for the comparison of power system expansion variants and export projects under uncertainty;

e) selection (using the risk matrix and game theoretic criteria) of the most effective export projects in the region. The conditions of such a selection depend on the form of the applied criterion (Laplace, Bayes, Savage, Hurwitz). In this paper, we consider the applicability of two criteria:

- Laplace criterion (average risk)

$$\overline{\lambda_r} = \min_i \frac{1}{N} \sum_{n=1}^{N} \lambda_i^n, \quad \text{at} \ \ i = 1,2,\dots,R \ .$$

- Savage criterion (minimax risk)

$$\overline{\overline{\lambda_r}} = \min_i \max_n \lambda_i^n, \quad \text{at } i = 1,2,\ldots,R \; ; \; n = 1,2,\ldots,N .$$

Joint application of formal game theoretic criteria in a general case does not ensure the choice of a single result, but reveals several projects virtually identical in terms of the applied criteria. Such projects are considered equally cost-effective. The existence of equally cost-effective projects causes principal impossibility of complete formalization of the decision making process under uncertainty and requires some other criteria to make the final decision.

4. A production and financial model is used to assess the commercial effectiveness of the export projects chosen at the previous stage, and the most preferable of them is selected (according to the assumed criterion of commercial effectiveness).

3 Results of the research

According to the proposed methodological approach, the paper presents the assessment of comparative effectiveness of four projects for the construction of export-oriented power plants and transmission lines from East Siberia to China:

Project 1 – Thermal power plant (TPP) on Kovykta gas in the south of the Irkutsk Region;

Project 2 – TPP on Kansk-Achinsk coal in the Krasnoyarsk Territory;

Project 3 –TPPs on coal in the Republic of Buryatia (Olon-Shibirskaya TPP) and Trans-baikal Territory (Novaya-Kharanorskaya TPP);

Project 4 – TPP on Mugunsky coal in the Irkutsk Region.

The main technical – economic indices of the export-oriented power plants and transmission lines are presented in Table 2.

The optimization linear model defining the electric power industry of East Siberia was applied to assess the system effectiveness of the considered projects. The electric power industry is represented by 6 regional power systems (Khakass, Tyva, Krasnoyarsk, Irkutsk, Buryatia and Transbaikal regional power systems). The implemented version of the model covers a period from 2014 to 2030.

The calculations were carried out for four variants of combinations of external conditions for the power industry development in East Siberia: low and high projected levels of power consumption in the region (195-245 billion kWh), low and high fuel prices for power plants: gas - 115-150 \$/tce, coal 40-75 \$/tce.

Calculations on the optimization model of electric power system expansion in East Siberia made it possible to determine an optimal mix of energy facilities necessary for the development of electric power industry in the region and related costs for each export project and for each of the four variants of combinations of external conditions.

The research shows that depending on the external conditions, the construction of export-oriented power plants and transmission lines can lead to a noticeable change in the structure of power generation and costs of power industry development in the region. Thus, either regional power system will export its surplus power, or export-oriented power plants will deliver their surplus power to the system.

Further, according to the proposed methodological approach, the payoff matrix of costs is constructed, and then it is used to construct the matrix of economic risks (Table 3).

The obtained matrix of risks is used to select export projects that are most preferable in terms of the game theoretic criteria assumed in the paper (Laplace, Savage).

The studies demonstrate that the joint application of formal game theoretic criteria makes it possible to identify two export projects that are effective in terms of the applied criteria.

Table 2. Technical and economic indices of export projects

Index	Projects

	1	2	3		4
	TPP on Kovykta gas	**TPP on Kansk-Achinsk coal**	**Olon-Shibirskaya TPP**	**Novaya-Kharanorskaya TPP**	**TPP on Mugun-sky coal**
Power plants					
Installed capacity, million kW	3.2	3.2	2.4	0.8	3.2
Type of equipment	CCP*	USCPP **	USCPP	USCPP	USCPP
Efficiency, %	55	45	45	45	45
Electricity generation, billion kWh	20.8	20.8	14.4	4.8	20.8
Specific fuel consumption, gce/kWh	222	276	276	276	276
Type of fuel	Gas	Brown coal	Hard coal	Brown coal	Brown coal
Specific capital investment***, $/kV	1000	1600	1500	1600	1600
DC transmission line					
Length , km	1100	1860	380		1100
Voltage, kV	500				
Specific capital investment:					
- linear part, thousand $/km	600				
- substations, million $/pcs	100				

Note: * - CCPP (a combined cycle power plant consisting of a 300MW steam turbine and two 279MW gas turbines); ** - USCPP (a power plant operating at ultra-super-critical steam parameters); *** - from here on in 2010 prices.

Table 3. Matrix of economic risks of the electric power system expansion in East Siberia, million $

Projects	Conditions for electric power system expansion				Criteria*	
	Variant 1	Variant 2	Variant 3	Variant 4	Laplace $\bar{\lambda}_r$	Savage $\bar{\bar{\lambda}}_r$
X_1	0	0	25.7	27.9	**13.4**	27.9
X_2	157	141	77	68	111	157
X_3	221	234	132	164	188	234
X_4	27.6	26.9	0	0	13.6	**27.6**

* in each column, the best variant is highlighted according to the respective criterion

In this case, the most preferable export projects are: *project 1,* that suggests the construction of a TPP on Kovykta gas as an export power plant, according to Laplace criterion (an average risk); and *project 4*, i.e. the construction of a TPP on Mugunsky coal, according to Savage criterion (a minimax risk).

Further, the simulation production and financial model is used to assess the commercial effectiveness of the chosen export projects, and the project to be implemented is selected. The preference is given to the project ensuring the minimum tariff of exported power at the sites where export transmission lines cross the border. The tariff is determined as a minimum power price acceptable for an exporter (disregarding excises and duties), which will compensate for operating costs (including payment for CO_2 emissions), investment costs, and ensure acceptable return on the invested capital.

The tariffs were calculated based on the following assumptions:

–design lifetime of power plant is 30 years for TPP on Kovykta gas, and 35 years for TPP on Mugunsky coal;

– design lifetime of an export ±500 kV transmission line is 25 years;

– internal rate of return is 15%;

– sources of funding are authorized capital;

Tax rates are assumed according to the current taxation system in the Russian Federation: tax on profit is 20 %;

– payback period for thermal power plants is 18 years;

– payback period for power transmission lines is 15 years;

– the projected price of Kovykta gas is 130-175 $/1000 m^3$, of Mugunsky coal - 25-30 $/t

Table 4 shows the export power tariffs (on the border with China) calculated using the simulation model for the assumptions made.

Table 4. Electricity tariff on the border with China

Index	**Measure-ment unit**	**TPP**	
		on Kovykta gas	**on Mugun-sky coal**
Tariff at the receiving end, total	cent/kWh	7.9-8.8	10.1-10.6
Including for:			
generation	cent/kWh	6.5-7.4	8.7-9.2
transmission	cent/kWh	1.4	1.4

According to Table 4, the project for construction of the export-oriented TPP on Kovykta gas (in the south of the Irkutsk Region) with power transmission by DC line to the border with China is more effective (according to the assumed criterion) compared to the project for construction of the TPP on Mugunsky coal in the Irkutsk Region. The export tariff on the border in the case of TPP on Kovykta gas is estimated at 8-9 cent/kWh, whereas for the TPP on Mugunsky coal - it is 10-11 cent/kWh.

4 Conclusion

1. A methodological approach is proposed to comprehensively assess the comparative effectiveness of power export projects under uncertainty of the power system expansion in the region.
2. The comparative effectiveness of the projects for construction of export-oriented power plants and transmission lines to supply power from East Siberia to China is assessed according to the developed methodological approach.
3. Consideration is given to four potential projects for large-scale power export from East Siberia to China, that suggest the construction of export-oriented power plants and DC transmission lines to the area of the town of Zabaikalsk (border with China): TPP on

Kovykta gas in the south of the Irkutsk Region; TPP on Kansk-Achinsk coal in the Krasnoyarsk Territory; TPPs on coal in the Republic of Buryatia and Transbaikal Territory (Olon-Shibirskaya TPP and Novaya Kharanorskaya TPP); and TPP on Mugunsky coal in the Irkutsk Region.

4. The studies show that the project for construction of export-oriented TPP on Kovykta gas with a DC power transmission line to the border with China (the town of Zabaikalsk) is most preferable compared to the other considered projects for construction of export power plants and transmission lines in East Siberia.
5. The proposed methodological approach and findings of the research provide additional information for the promotion of cooperation between Russia and neighboring countries (China, in particular) in the field of power export.

References

1. *Eastern vector of Russia's energy strategy: current state, look into the future*/ ed. by N.I.Voropai, B.G. Saneev. – Novosibirsk: Academic Publishing House "GEO". – 368 p. (2011) (in Russian)
2. *Effectiveness of interstate electric ties* /L.S. Belyaev, S.V.Podkovalnikov, V.A. Saveliev, L.Yu.Chudinova. – Novosibirsk: Nauka. – 239 p. (2008) (in Russian)
3. Likhachev V.L. *Shanghai Cooperation Organization and energy cooperation: current state and prospects for development* //Energy policy. No. **4-5**. P. 39-46. (2010) (in Russian)
4. *Potential and prospects for cooperation between China and Russia in the field of conventional and nonconventional energy* /Ed. by S.G.Luzyanin / Institute of Oriental Studies of RAS. – M.: Center of Strategic Conjuncture. – 254 p. Retrieved from: http://www.elibrary.ru/download/elibrary_2565091_64474007.pdf. (2014) (in Russian)
5. A.V.Lagerev, B.N.Khanaeva, K.S.Smirnov. *Priorities and prospects for the power industry development in East Siberia.* // Energetik. No.**8**. P. 2-7. (2011) (in Russian)
6. Makarov A.A., Melentiev L.A., *Methods for research and optimization of the energy sector.* – Novosibirsk: Nauka. - 376 p. (1973) (in Russian)
7. L.S.Belyaev. *Solving complex optimization problems under uncertainty.* – Novosibirsk: Nauka. – 328 p. (1978) (in Russian)
8. V.G.Sokolov, V.A.Smirnov. *Research into the flexibility and reliability of economic systems.* – Novosibirsk: Nauka. - 253 p. (1990) (in Russian)
9. S L.M.hevchuk, L A.S.ukianov, A.A.Kudryavtsev. *Risk analysis in the problems of strategic planning for large-scale companies*// Proceedings of RAS. Power Engineering. No.**2**. P. 52-64. (2000) (in Russian)
10. *Methods and models of forecast research into interrelations between energy and economy*/ Y.D.Kononov, E.V.Galperova, D.Y.Kononov, et al. – Novosibirsk: Nauka. – 178 p. (2009) (in Russian)
11. *Rationale of power system development: methodology, models, methods and their application*/ N.I.Voropai, S.V.Podkovalnikov, V.V.Trufanov, et al; Chief ed. N.I.Voropai. – Novosibirsk: Nauka. – 448 p. (2015) (in Russian)

Coal resources of the eastern regions of Russia for power plants of the Asian super ring

Aleksander Sokolov[1], and *Liudmila* Takaishvili[*,1]

[1] Melentiev Energy Systems Institute of Siberian Branch of the Russian Academy of Sciences, 130, Lermontov str., Irkutsk, 664033, Russia

Abstract. The eastern regions of Russia have a substantial potential for expansion of steaming coal production. The majority of coal deposits in the eastern regions are located close enough to the objects of the Asian super ring. The large coal reserves make it possible to consider it as a reliable fuel source for power plants for a long-term horizon. The coal reserves suitable for using at power plants of the Asian super ring are estimated in the paper by subject of the federation of the eastern regions for operating and new coal producers. The coal deposits of the eastern regions that are promising for the construction of power plants of the Asian super ring are presented. The paper describes both the coal deposits of the eastern regions that are considered in the projects for power plant construction and included in the program documents and the coal deposits that are not included in the program documents. The coal reserves of these deposits and the possible volumes of its production are estimated. The key qualitative coal characteristics of the deposits: heating value, and ash, sulfur, moisture content are presented. The mining-geological and hydrological conditions for deposit development are briefly characterized. The coals of the eastern regions are showed to contain valuable accompanying elements. It is noted that the creation of industrial clusters on the basis of the coal deposits is the most effective from the standpoints of the economy and ecology. The favorable and restraining factors in development of the described coal deposits are estimated.

1 Introduction

The eastern regions of Russia (East Siberia and the Far East) possess considerable coal reserves and resources. The majority of coal deposits in the eastern regions are located close enough to the objects of the project "The Asian super ring". According to the forecasts of U.S. Energy Information Administration [1] and International Energy Agency [2] the electricity demand in the Asian countries will rise and hence, the demand for energy resources will also increase. The coal reserves of the eastern regions are hard and brown coals and anthracite. The steaming coals form the main portion of the coal reserves (90%). The high-quality hard coals are attractive for export. The low-quality products of hard coal washing, the hard coals that do not comply with the quality requirements for export, and the brown coals are the fuel for power plants. The availability of substantial coal resources in the

* Corresponding author: luci@isem.irk.ru

eastern regions and provision with the coal reserves (for about 700 years) allow its consideration as a reliable fuel source for both the existing and new power plants for a long-term horizon. The steaming coal deposits of the eastern regions of Russia are located in the regions that differ in the development of the transport and social infrastructures, in the mining-geological, hydrological and other characteristics. The possibilities for the coal industry development in the eastern regions of Russia were studied by many authors [3, 4]. However, the possibilities for using these coals for power plants of the Asian super ring were not analyzed in these works. The studies of the coal industry development in the country and its regions at Melentiev Energy Systems Institute have been carried out for more than 40 years, which resulted in designing the system of models and the information system [5, 6]. The study of coal resources in the eastern regions of Russia for construction of power plants of the Asian super ring is seen as topical and possible.

2 Coal resources

The inferred coal resources of the eastern regions amount to 3184.9 billion t or 82% of the inferred coal resources in Russia (3927.7 billion t) [7]. The balance reserves of the eastern regions of categories $A+B+C_1$ (the reserves that are promising for development at the current stage of technological advance) make up 88 billion t (Table 1). In the future the large volumes of the coal reserves of category C_2 (39.4 billion t) and the off-balance reserves (28.4 billion t) can be developed on a commercial scale. The exploration level of the reserves (the ratio of the balance reserves to the inferred resources) is low and makes up only 2.8%. Hence, there are prospects for increasing the balance coal reserves. This can be achieved by performing the corresponding exploration works to include the inferred coal resources in the balance ones.

Table 1. Structure of balance coal resource of the eastern regions, billion t

Region / Type of coal	Balance reserves by category			
	$A+B+C_1$	C_2	Total $A+B+C_1+C_2$	Off-balance
Eastern regions, total	88.0	39.4	127.4	28.4
Brown	60.7	25.5	86.2	11.9
Hard	26.9	13.9	40.7	16.5
Including coking	9.0	4.5	13.5	1.4
Anthracite	0.03	0.03	0.05	0.02
Including for open-pit mining	71.3	28.0	99.3	15.2
Brown	57.7	22.6	80.3	10.3
Hard	13.6	5.4	19.0	4.9
Including coking	2.2	0.4	2.6	0.1

The balance coal reserves of the eastern regions contain primarily the reserves of steaming coals. The low-quality products of coking and steaming coal washing are energy resources. The major part of the coal reserves in the eastern regions (81% of the volumes of reserves) is suitable for open-pit mining. As to the brown coal reserves 95% of the reserves are appropriate for open-pit mining. The low-quality brown coals cannot be transported over long distances. They are used primarily for electricity production on the quarry site. The hard coals of the majority of the deposits in the eastern regions can also be used as a fuel for power plants.

3 Coal reserves of the eastern regions by subject of the federation

East Siberia and the Far East possess sufficient proved coal reserves for energy development, provision of domestic consumers, electricity export and construction of power plants, which potentially can be included in the Asian super ring. The steaming coal deposits of the Krasnoyarsk, Trans-Baikal and Primorye Territories, Irkutsk, Amur and Sakhalin Regions and the Republic of Buryatia are situated close enough to the objects of the Asian super ring (Fig.1).

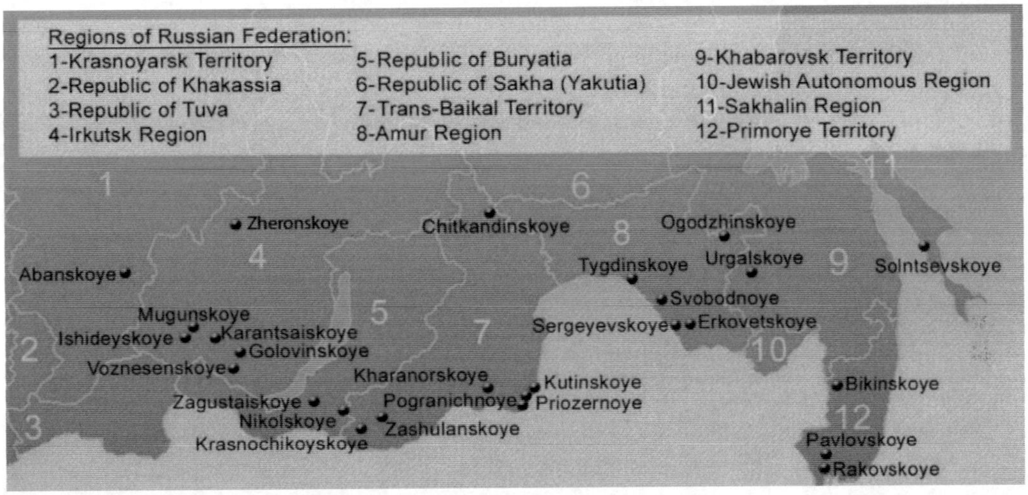

Fig. 1. Coal deposits for construction of power plants. [1]

The Krasnoyarsk Territory possesses substantial steaming coal reserves of categories $A+B+C_1$ – 45.5 billion t, category C_2 – 20.7 billion t. The Abanskoye coal deposit only is situated within the Asian super ring area, the size of its reserves exceeds the total size of the reserves of all remaining considered deposits.

In the Irkutsk Region the proved coal reserves of categories $A+B+C_1$ amount to 8.1 billion t, and those of category C_2 – 6.6 billion t. The structure of the reserves is characterized by the dominance of hard coals (73%), 7.7 billion t (96% of the proved reserves) are appropriate for open-pit mining. The total reserves of the prospective deposits make up 6.6 billion t. Coal production on these deposits can reach 43.5 million t, in particular, 18 million t on the existing open pits and 25.5 million t on the new open pits of the Mugunskoye and Golovinskoye deposits.

In the Republic of Buryatia, the proved coal reserves of categories $A+B+C_1$ amount to 2.2 billion t, and those of category C_2 – 0.3 billion t. In the structure of the reserves the share of brown coals equals 63%, 0.9 billion t (50% of the proved reserves) are suitable for open-pit mining. The total coal reserves of the prospective deposits make up 0.3 billion t. Owing to new construction the coal production on these deposits can reach 19 million tons.

In the Trans-Baikal Territory, the proved coal reserves of categories $A+B+C_1$ are 3.1 billion t, and those of category C_2 – 0.2 billion t. In the structure of the reserves the share of brown coals makes up 62%, 2.8 billion t (93% of the proved reserves) are appropriate for open-pit mining. The total coal reserves of the prospective deposits amount to 1.2 billion t. Coal production on the prospective deposits can reach up to 40 million t, including 20 million t on the existing Kharanorsky open pit.

[1] The map was compiled by an engineer of Melentiev Energy Systems Institute Muzuchuk R.I.

In the Khabarovsk Territory the proved coal reserves of categories A+B+C_1 amount to 1.6 billion t, and those of category C_2 – 0.7 billion t. The structure of the reserves is characterized by the dominance of hard coals (80%), 0.5 billion t (31% of the proved reserves) are suitable for open-pit mining. The Urgalskoye hard coal deposit only is situated within the Asian super ring area. Coal production can be increased on the existing mines and open pits.

In the Primorye Territory the proved coal reserves of categories A+B+C_1 amount to 2.3 billion t, and those of category C_2 – 1.4 billion t. The structure of the reserves is characterized by the dominance of brown coals – 2.1 billion t, 1.4 billion t (63% of the proved reserves) are suitable for open-pit mining. The total reserves of the prospective deposits reach 1.2 billion t. Coal production on the prospective deposits can amount to 9 million tons, including 7 million t on the operating Bikinskoye open pit.

In the Amur Region the proved coal reserves of categories A+B+C_1 amount to 3.6 billion t and those of category C_2 – 0.2 billion t. Brown coals form the bulk of the reserves – 3.5 billion t, which are appropriate for open-pit mining. The total coal reserves of the prospective deposits make up 3.5 billion t. Coal production on these deposits can reach 42 million t basically owing to new construction.

In the Sakhalin Region the proved coal reserves of categories A+B+C_1 amount to 1.8 billion t, and those of category C_2 – 0.6 billion t. The reserves of brown and hard coals are 0.95 and 0.85 billion t, respectively. Only 0.17 billion t are appropriate for open-pit mining. There are plans to increase coal production to 5 million t on the operating Solntsevskoye open pit and to 10 million t per year in the future. The key advantage of the coal deposits of the Sakhalin Region is their maximum proximity to the Asian super ring objects.

4 Prospective coal deposits for coal industry development

Currently the construction of large coal-fired power plants in East Siberia and the Far East, in particular, near the Asian super ring is studied in different state [8], federal and regional strategies and programs. Table 2 presents the list of the coal deposits, on which the construction of power plants with a capacity below 900 MW is considered in the program documents. It is intended to construct coal-fired power plants based on the run-of-mine coals and on the low-quality derivatives of coking and steaming coals.

The analysis of the coal resource potential in the East of Russia makes it possible to consider other coal deposits for construction of large power plants in addition to those enumerated in the Table 2 (Fig.1). They are: Abanskoye deposit in the Krasnoyarsk Territory; Golovinskoye, Voznesenskoye, Karantsaiskoye and Zheronskoye deposits in the Irkutsk Region; Svobodnoye, Sergeevskoye and Tygdinskoye deposits in the Amur Region. The available substantial balance coal reserves in the eastern regions contribute to the development of coal production owing to new construction and expansion of existing capacities. The coal deposits prospective for the coal industry development are presented in Table 3 [9-13].

On the majority of the presented deposits coal can be produced by the most advanced open-pit mining. The mining-geological and hydrological conditions for deposit development are of low and medium complexity. The thickness of coal beds that are prospective for development is 1.5 m and more, for individual deposits (Abanskoye in the Krasnoyarsk Territory) it can reach 25 m. The stripping ratios range from 2 m^3/t to 6 m^3/t. The coal deposits of the Irkutsk and Sakhalin Regions, the Primorye and Khabarovsk Territories, the southern areas of the Trans-Baikal Territory and the Amur Region are situated in the areas with the developed infrastructure. The deposits with the undeveloped infrastructure are characteristic of the northern areas of the federation subjects. The main difficulty in development of some deposits consists in climatic conditions and water content of the deposits, where the coefficient of the total moisture exceeds 30-40%.

Table 2. Coal deposits for construction of power plants

Deposit	Power plant capacity, MW
Erkovetskoye	4800
Kharanorskoye	2400
Urgalskoye	2400
Ishideyskoye	2000
Zashulanskoye, Krasnochikoyskoye	1500
Mugunskoye	900
Pogranichnoye, Priozernoye, Kutinskoye	900
Deposits of Trans-Baikal, Khabarovsk, Primorye Territories, Republic of Sakha (Yakutia), Amur and Sakhalin Regions	2900*

Source: [8]

Table 3. Description of coal deposits in the eastern regions

Deposit	Reserves, million t	Potential production, million t	Qualitative characteristics			
			Q^r, MJ / kg	A^d, %	S^d, %	W^r, %
Abanskoye	16766.8	35	14-16	8-15	0.3-0.8	31-38
Ishideyskoye	831.7	12	21-31	17-20	0.5-1.8	14-15
Mugunskoye	1639.3	13	17-20.1	16-21	0.5-1.6	18-20.6
Golovinskoye	186.0	5	19.7-23	13-23.1	1.9-2.8	14
Voznesenskoye	518.6	8	23-33	15-29	0.9-1.5	7.8-16
Kharantsaiskoye	3080.5	4	22	15	3.4	8.8-11.8
Zheronskoye	275.8	3	18-20	15.5-18	0.4-0.5	19.9-24.2
Nikolskoye	173.5	14.5	24	18	0.4-0.9	6
Zagustaiskoye	123.0	4.5	19	22	0.7	24
Kharanorskoye	742.4	15	12-13	5-25	0.3-0.7	33-44
Zashulanskoye, Krasnochikoyskoye	837.5	15-20	18-23	10-19	0.5-0.7	6-25/15
Pogranichnoye, Priozernoye, Kutinskoye	360	4.5	14.9-16.6	23-32	0.3-0.8	15-22
Chikandinskoye	12.5	1.2	22	17-23	0.1-0.4	1.5-7
Urgalskoye	1270.6	12	22.6-33.5	17-33	0.2-0.5	4.5-34
Pavlovskoye, Rakovskoye	463.9	2	11-12	16-33	0.3-0.5	41-44.5
Bikinskoye	908.9	7	8-24	18-35	0.4-1.0	34.44
Erkovetskoye	1040.2	18	12.3-13.4	15.4-18	0.3-0.6	35.7-38.2
Ogodzhinskoye	15.4	5	18.1-23.2	16.5-37	0.3-0.5	4.3-6.3
Svobodnoye, Sergeevskoye	1982.2	9	26.8-26.9	18.3-18.9	0.3-0.4	51-53.4
Tygdinskoye	466.3	10	7.3	20	0.3	57.2
Solntsevskoye	93.4	10	16-19.1	11.3-24.2	0.2-0.3	23

Note: Q^r – lower heating value; A^d- ash; S^d – sulfur; W^r– total moisture

Sources: [9-13]

The coals of the eastern regions comprise some valuable accompanying components [14-16]. The coals of the Ogodzhinskoye, Svobodnoye and Sergeevskoye deposits in the Amur Region include gold, sometimes in commercial concentrations [17]. The coals of the Erkovetskoye deposit contain platinum, the Bikinskoye deposit – germanium and accompanying elements, the Ishideyskoye deposit – silicon. The coals of the other deposits in the eastern regions also contain valuable accompanying components. These components can be extracted directly from coal, from wastes of coal mining, from ash and gases after coal combustion at energy facilities. The content of metals of some deposits exceeds the concentrations rational for their extraction many-fold, sometimes by dozens and even hundreds of times [18]. The coals also contain radionuclides, such as uranium, radon, thorium, radioisotopes of lead and potassium.

5 Conclusion

Availability of substantial coal reserves in the regions adjacent to the energy super ring being created can form a reliable base for fuel supply to the operating and new power plants.

The total balance coal reserves of the deposits prospective for the construction of power plants of the Asian super ring amount to 31.8 billion t. The potential production volumes of these deposits are estimated at above 180 million t, in particular, 60-65 million t owing to the increase in capacity and the reconstruction of the existing coal producers.

The favorable factors for the development of the prospective deposits are the following:

- availability of considerable coal reserves;
- relatively low or medium complexity of mining-geological and hydrological conditions for deposit development;
- location of the majority of the deposits in the areas with the developed infrastructure;
- possibility for the production of considerable coal volumes by the operating coal producers in the areas with the developed infrastructure.

The development of the majority of the prospective deposits can be restrained by the following factors:

- limitations of the existing transport infrastructure or its absence;
- difficulties in the development of individual deposits, particularly, in the Far East, due to the water content of the brown coal deposits;
- environmental restrictions on the development of coal production and coal-based generation.

Creation of industrial clusters on the basis of coal deposits that will unite the enterprises for production and conversion of coal and its man-made wastes, power plants on coal and gas from the coal beds, enterprises of coal chemistry and others is the most promising project from the standpoint of the economy and ecology [19]. The end products in similar projects are electricity and products with a high added value that are produced at coal conversion, degasification of coal beds, mine waters, coal dusts and emissions into the atmosphere after coal combustion at power plants. Extraction of valuable accompanying elements from coal is an important step of the transition to the clean combustion technologies.

The environmental aspects of the large-scale coal production expansion for construction of power plants of the Asian super ring need to be worked out thoroughly.

The choice of the sequence and priorities in development of the deposits requires the studies to be carried out on the efficiency assessment of deposit development and construction of power plants on the considered deposits.

References

1. *International Energy Outlook 2016* (the U.S. Energy Information Administration) Retrieved from: https://www.eia.gov/outlooks/ieo/, (2016)

2. *World Energy Outlook 2016* (International Energy Agency) Retrieved from: http://www.iea.org/newsroom/news/2016/november/world-energy-outlook-2016.html, (2016)
3. L.S.Plakitkina, Yu.A.Plakitkin, K.Diyachenko. *Prospects for coal production development in the Far-Eastern Federal District for the period until 2035 г.* // Mining journal, No. **3,** (2017) (in Russian)
4. M.I.Shchadov, O.A.Misevra. *The coal-energy balance of East Siberia and the Far East* (Mining industry journal) Retrieved from: https://www.litres.ru/m-schadov/ugolno-energeticheskiy-balans-vostochnoy-sibiri-i-dalnego-vostoka-25281703/, (2017) (in Russian)
5. G.V.Agafonov, A.D.Sokolov, L.N.Takaishvili. *Modeling of coal industry development*/ Proceedings of RAS. Power engineering journal. No.**6**, pp. 159-165. (2011) (in Russian)
6. A.D.Sokolov, L.N.Takaishvili. *Implementation of the software to forecast coal industry development in Russia's regions*. State-of-the-art technologies. System analysis. Modeling. No. **1(41)** pp. 126-133. (2014) (in Russian)
7. *State balance sheet of mineral reserves of the Russian Federation as of January 1, 2016.* Issue 91, Coal, I, Summary data. / M.: Ministry of Natural Resources and Environment of the Russian Federation, Federal Agency for Subsoil Management, Russian Federal Geological Foundation. - 382 p. (2016) (in Russian)
8. *Program of coal industry development in Russia for the period until 2030.* / Ministry of Energy of the Russian Federation, Moscow. -178 p. (2014) (in Russian)
9. *State balance sheet of mineral reserves of the Russian Federation as of January 1, 2016.* Issue 91, Coal, Volume VII, Siberian Federal District, part 2. – M.: Ministry of Natural Resources and Environment of the Russian Federation, Federal Agency for Subsoil Management, Russian Federal Geological Foundation. -314 p. (2016) (in Russian)
10. *State balance sheet of mineral reserves of the Russian Federation as of January 1, 2016.* Issue 91, Coal, Volume VIII, Far-Eastern Federal District. – M.: Ministry of Mineral Resources and Environment of the Russian Federation, Federal Agency for Subsoil Management, Russian Federal Geological Foundation, -378 p. (2016) (in Russian)
11. *Coal base of Russia. Volume III. Coal basins and deposits of East Siberia. Southern part (Kansk-Achinsk coal basin in the Krasnoyarsk Territory; Minusinsk basin in the Republic of Khakassia; Ulugkhem basin in the Republic of Tyva; Irkutsk basin and coal deposits of the Pre-Baikal area of the Irkutsk Region,).* – M.: CJSC "Geoinformmark". - 488 p. (2002) (in Russian)
12. *Coal base of Russia. Volume IV. Coal basins and deposits of East Siberia (Tunguska and Taimyr basins; deposits of the Trans-Baikal area).* – M.: CJSC "Geoinformmark". - 493 p. (2001) (in Russian)
13. *Coal base. Of Russia. Volume V. Book 1. Coal basins and deposits of the Far East (Khabarovsk Territory, Amur Reion, Primorye Territory, Jewish Autonomous Region).* – M.: CJSC "Geoinformmark". -371 p. (1997) (in Russian)
14. S.I.Arbuzov, V.S.Mashenkin, V.I.Rybalko, A.F.Sudyko. *Rare-metal potential of coals in Northern Asia (Siberia, Russian Far East, Kazakhstan, Mongolia)* // Geology and mineral resources of Siberia. No. **3**, pp. 41-44. (2014) (in Russian)
15. N. A.Lavrik. *Some premises of the integrated development of coal deposits in the south of the Far East*// Mining information-analytical newsletter. Separate issue: "Far East". pp. 420–430. (2005) (in Russian)

16. M.V.Golitsyn, V.I.Vyalov, A.Kh.Bogomolov, N.V.Pronina, E.Yu.Makarova, D.V.Mitronov, E.V.Kuzevanova, D.V.Makarov. *Prospects for development of the technological coal utilization in Russia.* Georesources. No. **2(61).** pp. 41-53. (2015) (in Russian)
17. *Decisions of the scientific-practical conference "Prospects for development of the coal chemistry in Russia: science, technology, production*"/ / Ministry of Energy of the Russian Federation, Kemerovo-14 p. (2016) (in Russian)
18. A.P.Verzhansky. *Ecologization of coal-based generation*. Coal. No.**9.** pp. 11-15. (2017) (in Russian)
19. Development program of the innovative territorial cluster "Integrated conversion of coal and man-made waste in the Kemerovo Region. Summary. - 46 p. (2012) (in Russian)

E3S Web of Conferences **27**, 02005 (2018) https://doi.org/10.1051/e3sconf/20182702005
AEC 2017

Comparative efficiency of technologies for conversion and transportation of energy resources of Russia's eastern regions to NEA countries

Aleksandr Kler, Elina Tyurina, Aleksandr Mednikov*

Melentiev Energy Systems Institute of Siberian Branch of the Russian Academy of Sciences, 130, Lermontov str., Irkutsk, 664033, Russia

Abstract. The paper presents perspective technologies for combined conversion of fossil fuels into synthetic liquid fuels and electricity. The comparative efficiency of various process flows of conversion and transportation of energy resources of Russia's east that are aimed at supplying electricity to remote consumers is presented. These also include process flows based on production of synthetic liquid fuel.

1 Introduction

There are considerable energy resources in the eastern regions of Russia, which make it possible both to meet domestic demand for energy carriers and to export them. However, the distances between regions that possess energy resources and potential consumption regions, first of all, in NEA countries, are great, which affects greatly the value of transportation costs. Therefore, the estimation of efficiency of long-distance transportation variants is getting topical. Since the share of gaseous and solid fuel in the energy balances is increasing the study on process flows of long-distance transportation of natural gas and coal energy, which include combination of technologies for conversion and transportation is of special interest [1-5].

Analysis of known technologies for conversion and transportation of energy carriers allows one to make the conclusion that the process flows including coal and gas conversion into synthetic liquid fuels (SLF) will be efficient for transmission of energy produced from gaseous and solid fuels at long distances. The benefit here will be achieved owing to the costs of liquid fuel transportation that are much lower than the costs of gaseous and solid fuel transportation. Besides, the costs of liquid fuel for consumers, first of all, power plants are much lower than those of coal consumers. At the same time, the process of SLF production is quite expensive. Therefore, it is necessary to estimate at what distance of transportation the additional SLF production costs will be recovered by saving the transportation costs [6-11].

* Corresponding author: tyurina@isem.irk.ru

The paper aims to compare the process flows based on technologies for SLF production and conventional technologies for energy transportation (pipeline transportation of natural gas, railway transportation of coal, electricity transportation over DC transmission lines). There are grounds to believe that at a large-scale long-distance transportation of coal, natural gas or SLF their marginal consumers will be power plants. Therefore it is sensible to assume electricity price for remote consumers, to be determined on the basis of value of internal rate of return, equal for all variants, as a criterion of process flow efficiency.

2 The plants for co-production of synfuels and electricity

Among technologies for large-scale production of various SLFs the most promising is the technology for production of methanol from synthesis gas which is in turn produced from coal or natural gas. This relates to the fact that methanol production is a well-known hetero-catalytic process: sufficiently selective, highly-productive and continuous. Methanol is suitable for transportation and storage [12]. The world's total methanol production capacity-accounts for about 45 million tons yearly.

It should be noted that methanol production is accompanied by release of a considerable amount of heat and formation of combustible blowdown gases. An efficient method for utilization of this "energy waste" is co-production of methanol and electricity at an integrated plant. These plants for co-production of synfuels and electricity are considered in this paper. Electricity produced by the plants is considered to be consumed in the area close to the plant. The price of this electricity is specified in the economic calculations.

The plants for co-production of synfuel and electricity are characterized by highly complicated flow charts and various physical and chemical processes that occur in their components. Besides, there is virtually no experience in their design. The research on the plants consists mainly of mathematical modeling and technical and economic studies on the models.

The works on mathematical modeling of the plants intended for synthesis of SLF and their technical and economic studies have been performed by the Energy Systems Institute of RAS for a long time [13-16]. Mathematical model of certain blocks of the plants on coal and natural gas, as well as the models of the whole plants were constructed. The studies made it possible to find optimal schemes and parameters, and identify the conditions for the plants to become economically efficient.

Fig.l shows as an example the calculated scheme of the plant for coal-based co-production of methanol and electricity. Conventionally the plant is represented by three blocks: block of synthesis gas production, block of catalytic synthesis of methanol and energy block.

In block I solid fuel is gasified and the mixture of hydrogen and carbon monoxide (synthesis gas) is produced. Besides, in this block synthesis gas is cooled in the system of gas generator heat exchangers and cleaned from ash and sulfur compounds, and CO2. Heat released in gas cooling is used for generation of steam that goes to steam turbine of energy block to produce electricity. In block II the catalytic synthesis of methanol is performed. This process is accompanied by generation of low pressure steam in intermediate heat exchangers of synthesis reactors that are intended for extraction of reaction heat. This steam goes to the low pressure section of the steam turbine. Blowdown gas goes from the block of synthesis to combustion chamber of gas turbine. The combustion products from the chamber are then used to generate electricity in gas turbines. In waste-heat boiler the steam of high and low pressure is generated. This steam goes to steam turbine for electricity generation.

It should be noted that the flow chart of the plant was developed on the basis of the most promising technologies of the processes that are employed in the plant. Fuel gasification takes place in gas generators with fluidized bed and dry slag removal, where steam-oxygen blast under the pressure of 2 MPa is used. This gas generator is an analogue of rather well studied and commercially implemented gas generator Winkler. Gas generators of the kind are most often applied in the plants for synthesis of SLF. The optimal flow chart of methanol synthesis with direct- flow reactors of synthesis and intermediate cooling of synthesis products in heat exchangers by boiling water was chosen based on the preliminary studies of various technologies: with isothermal reactors, with preliminarily switched synthesis reactor, with cooling of synthesis products by fresh synthesis gas, etc. The steam and gas combined cycle, being the most promising for energy plants, is suggested for energy block.

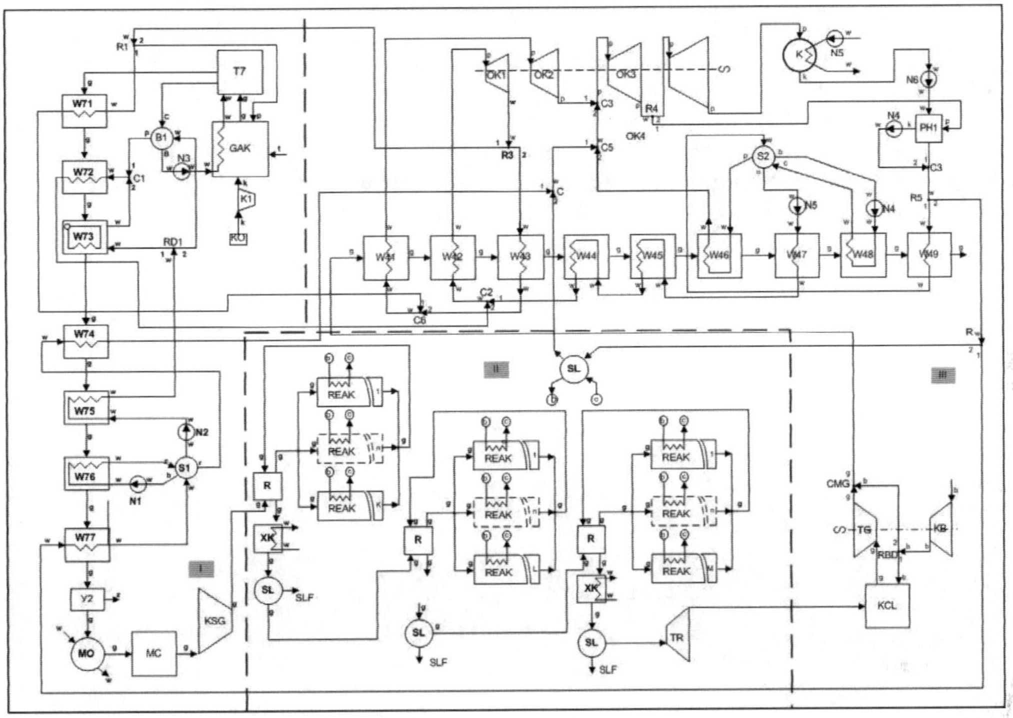

Fig. 1. Calculated scheme of a plant for coal-based methanol production: T7 – radiation heat exchanger on gasification products, GAK – gas generator, KO – block for oxygen production, Kl – oxygen compressor, W71–W77 – convective heat exchanger on gasification products, B1, S1, S2, S3 - separator drum, C, C1–C6 – mixer of water or steam flows, R1–R5 – separator of water or steam flow, N1–N6 – pump, RD1 – fractional separator of water or steam, Yl, Y2 – dry ash collector, MO – wet scrubber, MC – CO2 removal system, KSG – synthesis gas compressor, R – regenerative gas-gas heat exchanger, SL – separator, XK – condenser, REAK – adiabatic zone of SLF synthesis reactor, TP – expansion gas turbine. KCL – combustion chamber of gas turbine, TG – main gas turbine, KB – air compressor, RBD – air stream separator. CMG – mixer of gas and air streams, OK1–OK4 – steam turbine compartments, W41–W49 – convective gas-water heat exchangers of waste-heat boiler. K – steam turbine condenser, PHI – low pressure regenerative heater. Notation of flows: t – coal, k – oxygen, p – steam, w – water, steam, steam-water mixture, cooling water, g – conversion products, synthesis gas, blowdown gas, combustion products, b – air, kk – condensate, u, b, c – water supplied to circulating pump. 1,2 – bifurcation, mixing of flows. I – block of synthesis gas production, II – block of methanol synthesis. III – energy block.

The flow chart of the plant for co-production of methanol and electricity (PCME) on the basis of natural gas is similar to that presented above, except for the block for natural gas conversion, in which natural gas converter is used instead of gas generator on coal to produce synthesis gas.

Studies of the PCME aim to determine optimal parameters of the plants and change of their technical and economic performance depending on operation conditions (prices of fuel, equipment and products). The studies take into consideration the most important factor that influences the cost of all blocks of the plant and its energy efficiency, i.e. relationship between production of synthetic liquid fuel and electricity. The major parameters that determine this relationship include composition of blast supplied to gas generators on coal or natural gas converters, the number of reactors at synthesis stages, etc. These parameters were among the variables to be optimized in the study of the plant. Optimization of the plant included solving the problems of nonlinear mathematical programming which implies finding the plant parameters (composition of blast supplied to gas generators, catalyst volumes in reactors of synthesis, temperatures and pressures of working media of the combined cycle plant, etc.) that provide minimum value of the SLF price at set levels of internal rate of return (IRR) and prices of fuel consumed and electricity supplied, in terms of physical and technical constraints on the plant parameters.

Table 1 presents major technical and economic parameters of optimal variants of the plant for synthesis of methanol on the basis of coal and gas that were obtained from optimization studies on mathematical models of the plants.

Table 1. Main technical and economic parameters of optimal variants of PCME.

Parameter, unit	PCME variants	
	on gas	**on coal**
Annual consumption of natural fuel, million m^3, thousand tons	2200	4500
Annual consumption of reference fuel, thousand tce	2500	2500
Annual methanol production: - in reference fuel, thousand tce - in natural fuel, thousand tons	 1400 1900	 1350 1880
Capacity, MW: - steam turbine - gas turbine Auxiliary power Useful power	 330 160 167 320	 270 145 185 225
Annual electricity output, million kWh	2200	1560
Exergy efficiency. %	66,8	61,7
Capital investments in the plant, million USD	720	990
SLF price (IRR=15%), USD/tce	192	220

3 Process flows of electricity supply to remote consumers, on the basis of natural gas and coal energy

The considered plants for methanol synthesis were included in the process flows of electricity supply to remote consumers, on the basis of energy of natural gas and coal (brown coal) of large deposits.

The following process flows of energy production and transportation were compared (Fig. 2):

1. Natural gas production, 1420 mm gas pipeline, electricity production by gas combined -cycle plants (GP 1420).

2. Natural gas production, methanol production at PCME, methanol transportation by the 1220 mm pipeline, electricity production from methanol at combined cycle plant (PCMEg+PL 1120).

3. Coal production, methanol production at PCME, methanol transportation by the 1220 mm pipeline, methanol-based electricity production at combined cycle plant (PCMEc+PL 1220).

4. Coal production, railway transportation of coal from production site to consumers, coal-based electricity production by environmentally friendly thermal power plants (RWc+TPPc).

5. Coal production, electricity production from coal on production site by environmentally friendly TPPs, electricity transportation to consumers along DC transmission lines (TPPc+DCTL).

6. Natural gas production, electricity production from gas on production site by environmentally friendly TPPs, electricity transportation to consumers along DC transmission lines (TPPg+DCTL).

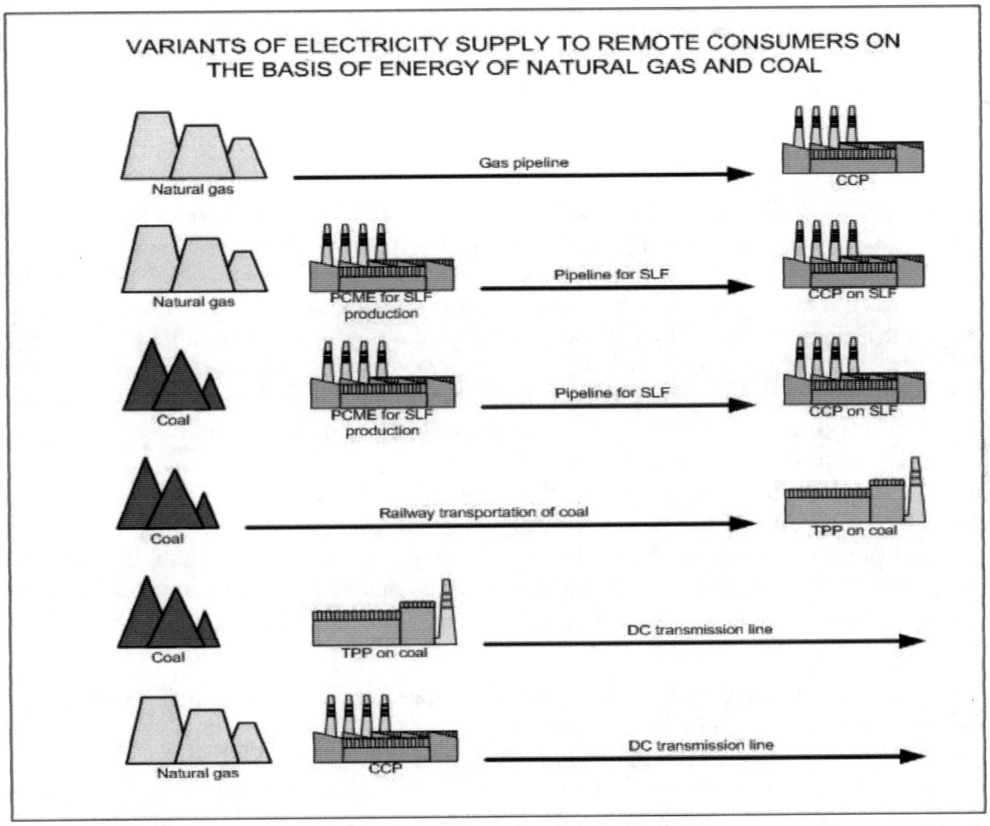

Fig. 2. The considered variants are characterized by the following features.

The economic efficiency of the indicated process flows is compared by the criterion of minimum electricity cost for consumers (at the given profitability level). A generalized statement of the optimization problem has the form:

$$\min_{C_{el}, C_t, IRR_z} C_{el} K^{TPP}, K^{PCME}, K^{tr}, U^{TPP}, U^{PCME}, U^{tr}, C_t), \quad (1)$$

subject to

$$H(x,y,s) \geq 0\,, \qquad (2)$$

$$G(x,y,s) \geq 0\,, \qquad (3)$$

$$IRR\left(K^{TPP}, K^{PCME}, K^{tr}, U^{TPP}, U^{PCME}, U^{tr}, C_t\right) = IRR_z\,, \qquad (4)$$

where C_{el} – electricity price at the process flow outlet; C_t – natural gas or coal price; x – a vector of independent variables of energy transport system; y – a vector of calculated parameters; s – a vector of initial technical and economic data; IRR_z – a set value of internal rate of return; H – a vector function of equality constraints; G – a vector function of inequality constraints; K^{TPP} – capital investments in electricity production; K^{PCME} – capital investment in conversion; K^{tr} – capital investment in transport; U^{TPP} – costs for electricity production; U^{PCME} – costs for conversion; U^{tr} – costs for transport depending on electricity supply variant.

Environmentally clean TPPs with the net efficiency 45% are used for electricity production from coal, combined-cycle plants with the net efficiency 52% are taken for electricity production from natural gas [17]. Natural gas and coal are converted into methanol at PCMEs. Electricity from methanol is produced at combined-cycle plants on methanol with the efficiency 55%. Natural gas is transported by the 1420 mm pipeline. The 1220 mm pipeline was suggested for methanol transport over long distances. Unit costs for methanol transport by the 1220 mm pipeline and natural gas transport by the 1420 mm pipeline were obtained by optimization calculations at the previous stages and amount to \$10/tce per thousand km for methanol and \$28/thousand m^3 per thousand km for gas pipelines [14-16]. Unit costs for railway coal transport are taken in accordance with the tariff directory of JSC "RZhD" [18]. They make up \$7.6–9.5/tons per thousand km. The price of coal in situ is taken equal to \$30/tce, that of gas – \$50/thousand m^3. The ±500 kV single circuit transmission lines with a rated capacity of 3000 MW were designed for DC transmission. Unit costs for electricity transmission by DC lines were also obtained at the previous stages of studies and make up 1 cent/kWh per thousand km. The main initial data applied in calculations are presented in Tables 2–4. The data concerning a plant for co-production of methanol and electricity are taken based on Table 1.

The results of comparing the economic efficiency of the mentioned process flows by the criterion of minimum electricity cost for consumers are given in Fig. 3.

Note that the 1220 mm methanol pipelines allow pumping of about 100 million tce. Such amount of fuel can produce some 400 billion kWh of electricity. Electricity transmission by the ±500 kV DC transmission lines of a rated capacity of 3000 MW allows delivery of about 20 billion kWh to consumers annually.

Table 2. Assumed technical and economic characteristics of dc transmission line.

Parameter, unit	**Value**
Specific investment in overhead TL, thousand doll./km	230
Specific investment in converter substation, doll./kW	85
Energy losses at converter substations, %	2
Total section of wires with the same pole, mm^2	2800

Annual energy losses in overhead line at rated output power and number of utilization hours, % /thousand km	3,75
Rated output power, MW	3000
Capacity utilization per year, h	6000
Capital investment in infrastructure, million doll.	80
Operating costs, % of investment – overhead line – converter substation	 0,8 2
Depreciation charges. % of investment	6

Table 3. Assumed indices of gas (GP) and methanol (MP) pipelines.

Parameter, unit	GP	MP
Rated pressure, MPa	7,4	5,4
Rated capacity utilization per year, h	8000	8000
Volume of pumped medium, gas – billion m^3. methanol – million tce	36	100.
Base unit cost of linear part, million doll./km	2,7	1,63
Fixed component of investment in pumping station, million doll.	25	15
Variable component of investment in pumping station, doll./kW	400	700
Specific fuel consumption by gas driven turbine at compressor stations of pipelines. gce/kWh	384	-
Electricity price of electric drive of pumping stations at methanol pipelines. cent/kWh	-	5
Rated low calorific value of natural gas or methanol. MJ/nm^3. MJ/kg	33,5	21,1
Operating costs. % of investment	3,5	3,5
Depreciation charges. % of investment	6	6
Cost of pumping, gas – doll./m3, methanol – doll./tce	28	10

Table 4. Assumed technical and economic indices of TPPs.

Parameter, unit	CCP on SLF	CCP on gas	TPP on coal
Specific investment per unit of useful power. doll./kW	700	700	1400
Specific consumption of equivalent fuel (net), tce/kWh	223	236	275
Conditionally fixed costs, % of investment	3–4	3–4	4–5
Depreciation charges, % of investment	4–6	4–6	5–7
Installed capacity utilization, h	6000	6000	6000

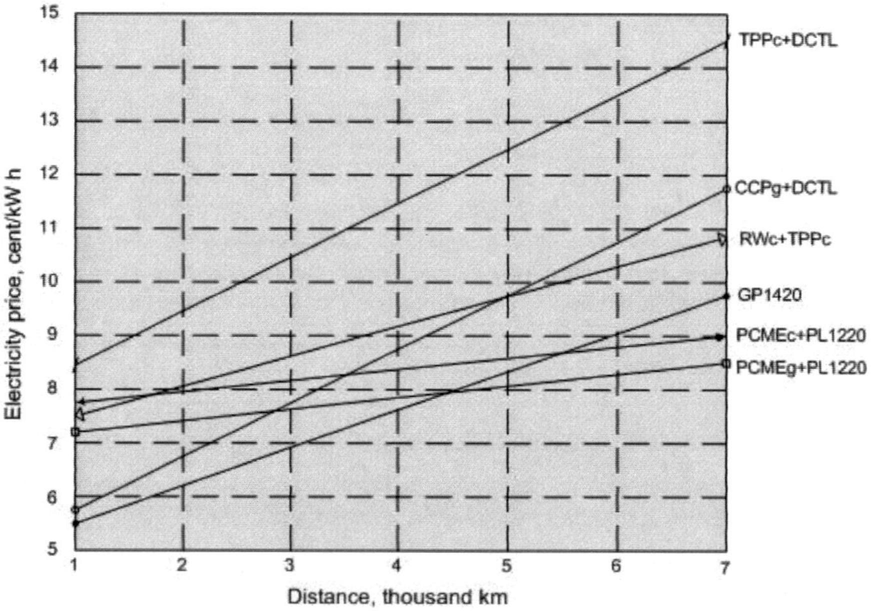

Fig. 3. Results of comparing the efficiency of variants for electricity supply to remote consumers.

4 Conclusions

Fig. 3 shows that transportation of natural gas by the 1420 mm pipeline is economically efficient over distances to 4,5 thousand km. For larger distances it is rational to use a PCME for natural gas conversion into SLF and transport it by a pipeline. Transport of (brown) coal by railway is reasonable over distances to about 1,7 thousand km. For larger distances it is advisable to transport SLF produced from coal by pipelines. Note that the tariff for coal transport by railway that is applied in calculations is accepted based on the current tariff policy of JSC "RZhD". The tariffs formed do not take into account capital investment in construction of new railways. It is expected that considerable volumes of coal transported can lead to limitation of capacity of operating railways and the necessity will arise to construct new branches. As a result the tariffs for coal transport will rise sizably.

Electricity transmission by DC transmission lines that is produced from brown coal or natural gas can not compete with distant piped large-scale transport of liquid and gaseous fuel.

Hence, it can be asserted with a high degree of validity that the process flows based on PCME for conversion of solid and gaseous fuel into SLF with its further piped transportation over large distances are most efficient as compared to other methods of distant transportation of energy of these fuel kinds.

Acknowledgements

This work was carried out at Melentiev Energy Systems Institute of Siberian Branch of the Russian Academy of Sciences and supported by Russian Science Foundation via grant № 16-19-10174.

References

1. Y. Xia, Y. Jin, K.P. Chen, M. Chen, D. Chen, J. of Nat. Gas Sci. and Eng., **41**, 112-124 (2017)
2. A. Vatansever, Energy Policy, **108**, 1-11 (2017)
3. W.Mazyan, A.Ahmadi, H.Ahmed, M.Hoorfar, J. of Nat. Gas Sci. and Eng., **30**, 487-514 (2016)
4. D.A. Wood, Ch. Nwaoha, B.F. Towler, J. of Nat. Gas Sci. and Eng., **9**, 196-208 (2012)
5. D. Osborne, *The Coal Handbook: Towards Cleaner Production* (WPL, 2013)
6. S. Yang, Z. Liu, Z. Tang, Yi. Wang, Q. Chen, Yu. Sun, Energy Conve. and Manage., **150**, 375-381 (2017)
7. A. Bassani, G. Bozzano, C. Pirola, E. Ranzi, S. Pierucci, F. Manenti, Energy Procedia, **105**, 4519-4524 (2017)
8. A. Śliwińska, D. Burchart-Korol, A.Smoliński, Sci. Total Environ., **574**, 1571-1579 (2017)
9. H.Huang, H.Xiao, S. Yang, App. Therm. Eng., **125**, 891-903 (2017)
10. T. Blumberg, T. Morosuk, G. Tsatsaronis, Energy. (to be published)
11. X. Liu, J. Liang, D. Xiang, S. Yang, Y. Qian, J. of Cleaner Production, **129**, 53-64 (2016)
12. C. Zhang, K. Jun, G. Kwak, Y. Lee, H. Park, J. of CO2 Utilization, **16**, 1-7 (2016)
13. A.M. Kler, E.A. Tyurina, *Mathematical modeling and technical and economic studies on plants for coproduction of methanol and electricity for methanol synthesis.* (Novosibirsk, Nauka. 1998) (In Russian)
14. A.M. Kler, B.G. Saneev, A.D. Sokolov, E.A. Tyurina, J. Energetika, **2**, 36-43 (2000) (In Russian)
15. A.M. Kler, B.G. Saneev, A.D. Sokolov, E.A. Tyurina, J. Energy perspectives, **6**, 53-60 (2002) (In Russian)
16. A.M. Kler, E.A. Tyurina, J. Energy perspectives, **9**, 69-78 (2005) (In Russian)
17. Reference calculation of transportation cost. http://rpp.rzd.ru/Rzd/
18. O.N. Favorsky., V.L. Polishchuk, I.M. Livshits, V.I. Dlugoselsky, J. Therm. Eng., **9**, 46-51 (2007)

The Russian-Mongolian vector of the eastern energy strategy of Russia: role of energy potential of the Baikal region

Boris Saneev[*,1], *Irina* Ivanova[1], *Alexander* Izbuldin[1], *Svetlana* Muzychuk[1], *Elena* Maysyuk[1], *Gennady* Borisov[2], and *Sodovin* Butkhuyag[3]

[1] Melentiev Energy Systems Institute of Siberian Branch of the Russian Academy of Sciences, 130, Lermontov str., Irkutsk, 664033, Russia
[2] Department of Regional Economic Research BSC SB RAS, Ulan-Ude, Russia
[3] Mongolian State University of Science & Technology, Ulan-Bator, Mongolia

Abstract The paper is concerned with the specific features of Russia's economic development in a new economic environment that caused the need to revise the priorities of energy policy. The research presents the initial conditions, targets and strategic directions of energy development in the East of the country. The focus is made on the priority lines of innovation and technology cooperation between Russia and Northeast Asian countries in the field of energy, and recommendations on necessary conditions and initiatives for their successful implementation are given.
Keywords: power export, gas main, consumer gasification, supplies of oil products, inter-country power system, renewable energy sources.

Introduction

The national interests of Russia require intensification of the mutually beneficial cooperation with the countries of Northeast Asia. Great significance in the entry of the Russian energy resources to the markets of Mongolia, Japan, China and Korea is attached to East Siberia and the Far East. Among the eastern regions of the Russian Federation, the Baikal region including the Irkutsk Region, the Republic of Buryatia, and Transbaikal Territory occupies a special place for Russia to enter the dynamically developing market of Mongolia and is an outpost in the implementation of the mutually beneficial energy cooperation between the two countries.

The role of NEA countries in the world economy is increasing, and therefore, this region is getting more significant as a contact zone of Russia with Mongolia and China. Furthermore, since some ports in the Black and Baltic seas were lost for Russia, the role of transportation infrastructure of the region in the performance of its transit functions in order to service the domestic and international economic and energy relations is increasing

* Corresponding author: saneev@isem.irk.ru

Current state and potential of energy industries in the region

A large fuel and energy basis of the country is established in the Baikal region. In 2015 it produced more than 34 million t of coal, generated about 62 billion kWh of electricity, and refined 9.1million t of oil. Hydropower plants of the Baikal region generated nearly 17% of all electric power in the Siberian Federal District [1].

The energy sector of the Baikal region makes a pronounced contribution to production indices of the country. In 2015 its enterprises produced 9.2% of Russian coal, generated 5.8% of electricity, refined 3.2% of crude oil and 3% of oil. The regional energy sector covers basically demands of the own economy for energy resources. In 2015 its share in consumed coal in the country was 12.1%, electricity – 6.2%, thermal energy – 4.3%, oil products – 3.9% [1].

Coal industry is of great importance for the economy of the Baikal region as it supplies fuel to the enterprises of electric power and heat power industries, production facilities, and population. Coal is the major fuel in the region.

Geological resources of coal in the territory of the region make up 28.9 billion t (32 % of national), balance reserves – 29.5 billion t as of 01.01.2015 [6]. Moreover, 87 % of the total reserves are concentrated in the deposits with favorable development conditions. This allows open-cast mining, which is the most effective way of coal production.

Electric power industry of the Baikal region is a key industry. It is represented by 36 thermal and hydro power plants. The installed capacity of the power plants in the Baikal region makes up nearly 16.5 GW, of which 55% falls on hydro power plants (above 9 GW). Four hydro power plants that operate in the region produce above 58% of electricity. Moreover, the region has nearly 500 diesel power plants with a total capacity of 420 MW, which are used as backup and autonomous energy sources [1].

Heat in the region is produced by thermal power plants, numerous boiler plants, heat utilization plants and electric boiler plants.

The level of gas use in the Baikal region is extremely low – nearly 5%, whereas for Russia on average it is above 66.2%, and in the western regions of the country even higher [3]. At the same time the explored and prepared gas reserves are sufficient to begin the commercial development of the fields and convert consumers to gas.

The Baikal region has unique reserves of hydrocarbon resources, which are mainly concentrated in the territory of the Irkutsk Region. As of 01.01.2015 the state balance registered 37 fields of hydrocarbon fuel with a total recoverable reserves of: oil – 907.2 million t, natural gas – 4.7 trillion m^3, condensate – 182.2 million t in the region [4, 5]. The field unique in reserves (categories C_1+C_2 about 2.5 trillion m^3) is Kovykta gas and condensate field. There is also Levoberezhnoe field, Angaro-Lenskoe, Chikanskoe, Bratskoye and other fields.

Oil production in the Irkutsk oblast reached 15.8 million t in 2015 that is supplied to the oil pipeline system "East Siberia – Pacific ocean" (ESPO) basically for export. N*atural gas production* increased to 2.2 billion m^3 in 2015. However, it is used extremely inefficiently, mainly it is burnt at flare units, and pumped back into the reservoir [5]. Only insignificant part of the gas produced is consumed for auxiliary needs of the gas producer facilities.

Oil refining in the territory of the Baikal region is performed by the Angarsk Petrochemical Company with a refining capacity of 10.2 million t of crude oil a year. Owing to a considerable share of secondary processes, the Angarsk oil refinery is characterized by a higher oil refining ratio – 76.5 %. In 2015 the company refined 9.1 million t of oil, produced 8.11 million t of marketable product [6].

 https://doi.org/10.1051/e3sconf/20182702006

The Angarsk oil refinery plays an important part in supply of oil products to the consumers in Siberia and the Far East. The refinery practically completely provides consumers in Baikal region. Moreover, its products are sold outside the region and are exported to 14 countries of the world: China, Mongolia and other NEA countries. The share of export in the total amount of supplies makes up 26 %.

The territory of the Baikal region is characterized by a considerable potential of *renewable natural energy resources*. There is 37% of hydro energy, 43% of solar energy, 30% of wind energy and more than 50% of wood biomass potential in East Siberia. The most favorable conditions for the use of solar potential for energy purposes are in the southern areas of the region and those of wind potential – in the depression of Lake Baikal (Fig.1) [7].

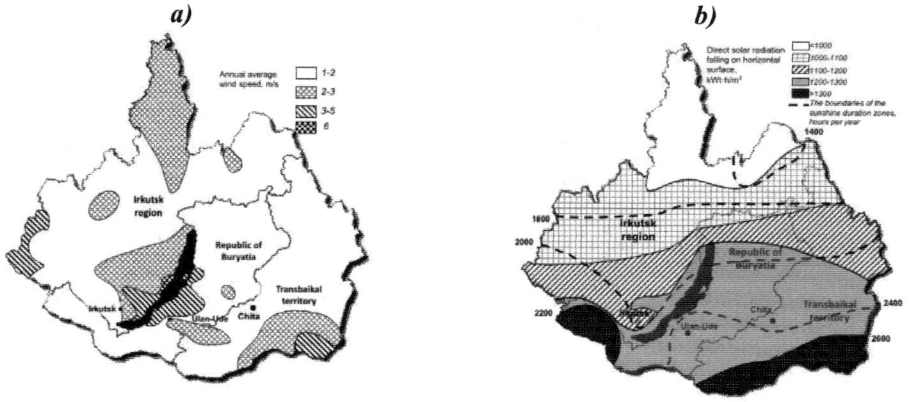

Fig. 1. Zoning of the Baikal region territory by the indices: a) wind potential, b) solar potential.

State and strategic directions for the bilateral energy cooperation

Currently Russia is ranked third in the amount of Mongolia's export and first in the amount of its import. And more than half the Russian-Mongolian commodity turnover falls on the Irkutsk oblast [8]. Today the key part in the development of mining industry of Mongolia is played by the Mongolian-Russian company "Erdenet". Russia annually supplies to Mongolia 140-150 million kWh of electric energy, and 560-580 thousand t of oil products. Russian oil products provide for 92 % of Mongolia's demand. In 1999 Russia started to supply liquefied natural gas.

In the Declaration on development of the strategic partnership between the two countries the arrangements were reached on development of cooperation in different fields, including the improvement of transportation infrastructure, joined development of mineral deposits, production of uranium, modernization of joint ventures: Erdenet Mining Corporation, Mongolrostsvetmet LLC and JSC "Ulan Bator Railway", and development of cross border interaction.

The focal issue of the cross border cooperation is the creation of a special economic zone in the area of the towns Kyakhta and Altanbulag as well as the development of infrastructure and engineering communications, first of all railway and automobile roads.

The prospects for the cooperation between the two countries in the energy field are related to the implementation of strategic priorities in the development of oil industry, gas industry, coal industry, power industry and renewable energy.

An important aspect of oil and gas cooperation between the two countries is expansion and specification of the resource base, related to the geological surveys to be conducted, increase in the proved reserves, additional exploration and discovery of hydrocarbon fields in the territory of Mongolia.

 https://doi.org/10.1051/e3sconf/20182702006

The Mongolian party is interested in further increase in the supplies of oil products to Mongolia's market by the Russian companies to 1.5 million t In this context, there are intentions to extend acting contracts concluded with the JSC "NK Rosneft". Currently there is an interest of the JSC "NK Rosneft" in further strengthening of energy cooperation and supply of oil products to the small-scale wholesale and retail segments of Mongolia's market through the creation of a network of refueling stations.

Taking into account the construction of the Darkhan oil refinery in Mongolia, it is necessary to consider the possibility of supplying oil either by oil pipeline system to be constructed, or by railway.

Prospective oil and gas cooperation between Russia and Mongolia is also related to the possibility of creating a joint venture on production and mutually beneficial distribution of products. The parties support the idea of stable supplies of liquefied hydrocarbon gas from Russia to Mongolia on mutually beneficial terms.

Furthermore, there are preconditions for supply of network gas to consumers in the Baikal region and Mongolia in the event that the export gas pipeline "Russia-China" is constructed in their territory. Based on the available estimates [9], large gas consumers are concentrated in the cities of Sukhe-Bator, Darkhan and Ulan-Bator. Their annual demand for natural gas is estimated at 0.4-0.6 billion m^3.

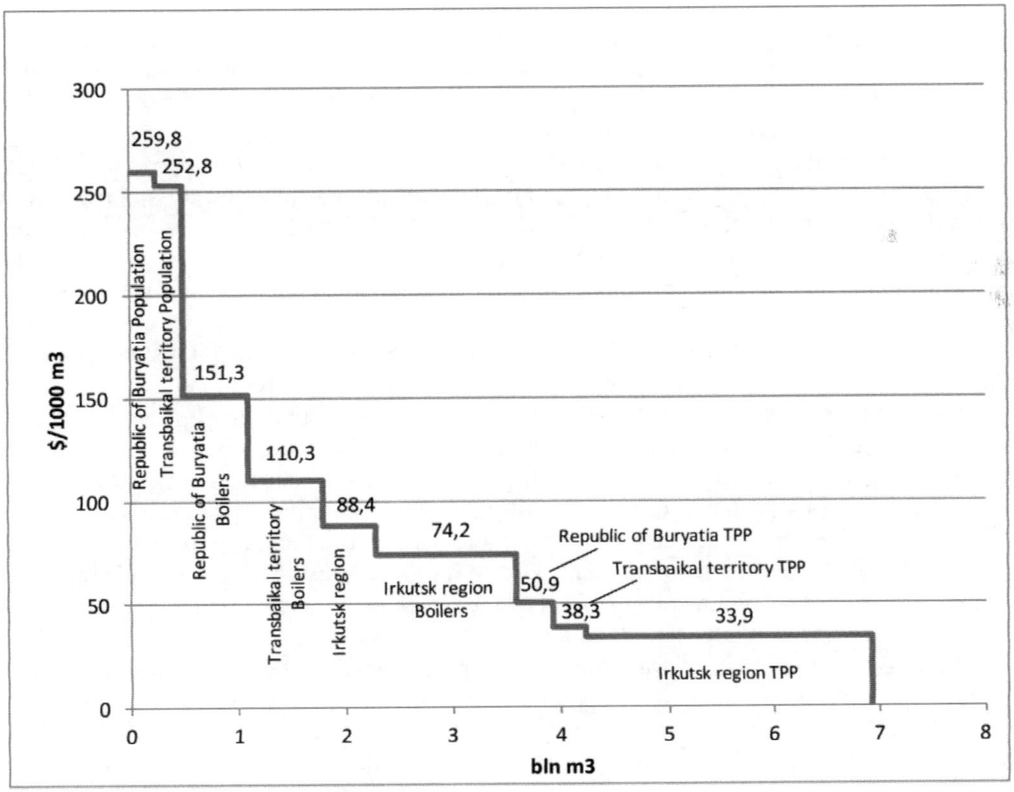

Notes. Initial data for calculations: Dollar exchange rate – 60 RUB/doll.

Fig. 2. Competitive prices of natural gas and potential gas demand for the population, boilers and thermal power plants of the Baikal region.

Since there is interfuel competition in different categories of consumers, the maximum demand for gas in the region is estimated at 7-8 billion m^3 by 2030, including: 4-4.5 – Irkutsk Region; 1.0-1.5 – Republic of Buryatia; 1.0-1.5 billion m^3 – Transbaikal Territory (Fig.2) [10]. The Table presents initial indices to determine competitive natural gas prices for thermal power plants, boiler plants, and population of the RF entities located in the Baikal region.

Table. Cost indices as of July 2017

	Electricity prices for population, RUB/kW·h	Coal prices, RUB/t	
		Boilers	Thermal power plants
Transbaikal Territory	2.89	1595	1131
Republic of Buryatia	2.97	2187	1500
Irkutsk Region	1.01	1073	588

Construction of a special gas pipeline system for supplying gas to consumers in the Republic of Buryatia, Transbaikal Territory and Mongolia with an annual gas consumption of 1.7-2.1 billion m^3 is commercially inefficient. This variant can be implemented only in the case of financial support from Russia and Mongolia. The supply of Irkutsk natural gas to Mongolia is cost-effective only if there is a large-diameter main export gas pipeline laid in these territories [11].

In the coal industry Russia and Mongolia can cooperate in the project for the development of the world's largest Tavantolgoi coking coal deposit.

Among the main directions in the cooperation between Russia and Mongolia in this field are the development of infrastructure and engineering communications, first of all railways, which make it possible to connect coal producers to consumers, and supply up-to-date mining equipment. Russian JSC "INTER RAO UES" is ready to take part in the construction of an infrastructural thermal power plant on the deposit.

Construction of a railway 1.1 thousand km long from the Tavantolgoi deposit to Russia's border is one of the options for development of the transport corridors for possible export of its coking coke to the market of the NEA countries [12]. JSC "Russian Railways" is willing to participate in the project and thereby combine construction of the railway with the development of Mongolian mineral resources, which will substantially increase coal transportation by rail.

The cooperation in *electric power industry* is connected to two directions: construction and reconstruction of power plants and creation of an interstate power system.

One of the problems in the Mongolian energy sector is the shortage of peak electric and thermal power during winter peak loads in Ulan-Bator. This problem can be solved by implementing different energy projects [13]. For instance, the construction of a peak thermal power plant in the east of Ulan-Bator that will run on gas fuel exported from Russia.

At the 20th meeting of the Russian-Mongolian Intergovernmental Commission on Trade-Economic and Scientific-Technological Cooperation the Russian party informed the Mongolian party of the technical capability of increasing electricity supplies from Russia to Mongolia and expressed interest in their augmentation [14].

The Russian party informed the Mongolian party of the interest of JSC "EuroSibEnergo" to develop the cooperation with Mongolian companies in the electric power sphere, in particular in establishment of joint enterprises, performance of special construction and erection and commissioning works at power plants in the territory of Mongolia, delivery of electric power equipment, performance of work and provision of consulting services on the power quality analysis and control, staff training and its professional advancement.

One of the promising directions in the cooperation between Russia and Mongolia (and other NEA countries) is the project intended for the creation of an interstate electric power system, which implies the development of electric ties between the eastern regions of Russia and neighboring NEA countries. The cross-border transmission lines constructed for this purpose can be used both to export electricity from Russia and to integrate electric power systems in the indicated countries for their joint (or simultaneous) operation and formation of common electricity and capacity markets [15].

At present, the interstate electric ties between Russia and Mongolia are represented by two transmission lines: a 220 kV transmission line from the Gusinoozerskaya condensing power plant to Darkhan and a 110 kV transmission line from the Kharanorskaya condensing power plant to Choibalsan. Further increase in electricity export will inevitably require both installation of suitable transmission lines, and commissioning of additional generating capacities.

The implementation of this scheme helps solve power supply reliability problems in a number of entities of the Russian Federation and opens up a possibility of parallel operation of IPS of Siberia and IPS of East, and electricity export to Mongolia and China.

The need for such an interstate interconnection is also recognized by the Mongolian party, since it is supposed that in the nearest future almost all power systems in Mongolia will experience power shortage of 400 MW in total.

According to the Memorandum between the Ministry of Energy of the Russian Federation and the Ministry of Energy and Mineral Resources of Mongolia dated October 14, 2010, the relevant departments provide necessary assistance to the companies of both countries in electric power cooperation to enable parallel operation of the Western power system of Mongolia and the power system of Russia.

Implementation of "The program for creation of an economic corridor Russia-Mongolia-China" signed at the 3rd meeting of the heads of the states dated 23.06.2016 in Tashkent makes an essential contribution to development of the Russian-Mongolian cooperation, in particular in the energy sphere. The program comprises 32 projects of cooperation and only one energy project. The following projects can be considered as potential: construction of the gas pipeline "Irkutsk – Ulan-Ude – Ulan-Bator – China (on the basis of the Kovykta field), construction of the 500 kV transmission line "Gusinoozersk – Ulan-Bator, construction of the Mokskaya HPP on the Vitim river and the Ust-Kutskaya TPP on the associated gas of the fields in the North of the Irkutsk oblast [16].

The promising direction in cooperation between Russia and Mongolia is improvement of energy effectiveness and development of *renewable energy.* The Mongolian part expressed its interest in this area at the 20th Meeting of the Russian-Mongolian Intergovernmental Commission on Trade-Economic and Scientific-Technological Cooperation and proposed consideration of the possibility for exchanging experience in solving these problems [14]. Recently the issues of renewable energy development in Mongolia have been devoted much attention, sets of projects on construction of solar, wind and hydro power plants have been under consideration [17].

Participation of the Russian specialists in the development and feasibility studies of joint projects for construction of small-scale hydro power plants could be one of the directions of cooperation in this sphere. The experience and knowledge of the Russian specialists will make it possible to ensure delivery of equipment for small-scale hydro power plants and provide professional training in its maintenance for the staff. An important joint project within this direction is the construction of renewable energy plants in the Baikal-Khubsugul zone in order to mitigate an anthropogenic impact of energy facilities on the environment.

Conclusions

1. A need has arisen to devise a scientifically grounded strategy for energy cooperation

between Russia and Mongolia under the conditions of energy integration and collaboration with the NEA countries. The strategy should reflect the order of development of energy resources, show the sequence and stages of energy supply to domestic consumers and energy export, and assess the socioeconomic consequences of the decisions made not only for the companies, but for the regions and countries as well.

2. At present, the main direction of long-term energy cooperation between Russia and Mongolia is determined and the raw material base in both countries is well studied [18, 19].

3. More attention should be paid to the coordinated measures taken by the participants (countries, regions, companies), i.e. to the economic, legislative, and other initiatives connected to the implementation of large-scale interstate energy projects. Due consideration should be given to the energy pricing formulas.

4. Research and design energy institutes, specialized higher educational institutions, scientific societies, and energy companies in Russia and Mongolia should take an active part in achieving these goals, and make a valuable contribution to solving the problem, which is important both for these two countries and for the whole Northeast Asia.

References

1. Forms of statistical reporting on the energy sector. M.: Rosstat (2015).
2. State balance of mineral reserves of the Russian Federation. Coal. Volume VII, Siberian Federal District. Part 2. M.: Ministry of natural resources and ecology of RF. (2016), 314 p. (2016) (in Russian)
3. Data of PJSC "Gazprom". Gasification. URL:http://www.gazprom.ru/about/production/gasification/ (in Russian)
4. Data of PJSC "Gazprom". Reserves of the Kovykta gas condensate field. URL:http://www.gazprom.ru/about/production/projects/deposits/kovykta/ (in Russian)
5. State and utilization of mineral resources of the Irkutsk oblast (as of 01.01.2015) URL:http://atlaspacket.vsegei.ru/_Files/пакеты_05_2016/SFO/PDF/IRKUT_OBL/31_32_spravka_msb.PDF (in Russian)
6. Data of PJSC "NK "Rosneft" URL:https://aoanhk.rosneft.ru/about/Glance/OperationalStructure/Pererabotka/aoanhk/Osnovnie_pokazateli/ (in Russian)
7. Energy sector of the Baikal region: current state, prospects for development / Ed. by B.G.Saneev, Russian Academy of Sciences. Siberian Branch, Melentiev Energy Systems Institute. Novosibirsk: Academic Publishing House "GEO", 176 p. (2015) (in Russian)
8. Chudanzhiy Sharav. Current state of economic reforms in Mongolia. Izvestiya of Irkutsk State Economics Academy, No. 5 (55), P.110-113 (2007) (in Russian).
9. A long-term Vision of Natural Gas Infrastructure in Northeast Asia, 2005 Year Version. Northeast Asian Gas & Pipeline Forum, (September 2005), Tokyo, Japan, 94 p.
10. B.G.Saneev, L.A.Platonov, E.P.Maysyuk, A.K.Izhbuldin. Multipurpose utilization of natural gas in the Baikal region: prerequisites, directions, conditions for implementation. Region: economy and sociology, No. 3, P. 190-202. (2012) (in Russian)
11. Eastern vector of Russian energy strategy: current state and future projections / Ed. by N.I. Voropai, B.G. Saneev, Novosibirsk: GEO, 368 p. (2011) (in Russian).
12. Gotov Dugerjav. Enhancing Northeast Asia and Mongolia cooperation through transport network development [Electronic resource]. - http://www.kiep.go.kr/include/filedown.jsp?fpath=POOL0206&NO=185518&FNO=962 (accessed 16.10.2017).

13. H.Enkhzhargal, S.Batmunkh, V.A.Stennikov. Prospective lines of power system development in Mongolia. Energy policy, Issue 4, P. 70-81. (2012) (in Russian)
14. Minutes of the 20th meeting of the Russian-Mongolian Intergovernmental Commission on Trade-Economic and Scientific-Technological Cooperation dated 13, Ulan-Bator, 20 p. (December, 2016) (in Russian)
15. G.O.Borisov Problems and prospects for energy development of Zabaikalsky krai. Materials of the round-table discussion no. 2 “Development of energy and energy infrastructure as a factor of economic growth in Siberia and the Far East”, International Economic Conference “New economy – new approaches”, Ulan-Ude (2012), Electronic resource: http://egov-buryatia.ru/index.php?id=4948. (in Russian)
16. G.O.Borisov, Z. B.-D.Dondokov, V.O.Namzhilova. Economic corridor: China-Mongolia-Russia: wait mode. ECO, No. 5 (515), P.98-106 (2017) (in Russian)
17. National policies for new and renewable energy of the countries in North-East Asia region. Country Reports of the countries in Northeast Asia. Seoul: Korea Energy Economics Institute, P. 15-40. (2016)
18. N.I.Voropai, B.G.Saneev, S.Batkhuyag, Kh.Enkhzhargal. Energy cooperation between Mongolia and Russia: current state and strategic directions (results of the Russian-Mongolian project of 2011-2012). Spatial Economics, No. 3, P.108-122. (2013) (in Russian).
19. N.I.Voropai, B.G.Saneev, I.Yu.Ivanova, A.D.Sokolov, E.P.Maysyuk, N.M.Sysoeva, G.O.Borosov, S.Batkhuyag, Kh.Enkhzhargal. Forecasting of strategic directions in energy cooperation between Russia and Mongolia. Scientific cooperation of the Siberian Branch of RAS with Mongolia in the first half of the 21st century. Issue 1: 2011-2013. Novosibirsk: Publishing House SB RAS, P.123-140. (2015) (in Russian)

AUTHOR INDEX